New And Efficient Approach And Closed-Form Confidence Intervals for Parameters of
Normal, Exponential and Gamma Distributions

By Vincent A. R. Camara, Ph.D.

Table of contents

Chapter 1 Introduction

1.1 Bayesian philosophy

1.2 Aims and Objectives

Chapter 2 Approximate Bayesian Confidence Intervals for the Variance of a Gaussian distribution corresponding to the Square Error Loss function.

2.1 Introduction

2.2 Preliminaries

2.3 Main Results

2.4 Numerical results

2.5 Summary and conclusions

Chapter 3 Approximate Bayesian Confidence Intervals for the Variance of a Gaussian distribution corresponding to The Higgins-Tsokos loss function.

3.1 Introduction

3.2 Preliminaries

3.3 Main Results

3.4 Numerical results

3.5 Summary and conclusions

Chapter 4   Approximate Bayesian Confidence Intervals for the Mean of a Gaussian distribution corresponding to the Square Error Loss function.

4.1 Introduction

4.2 Preliminaries

4.3 Main Results

4.4 Numerical results

4.5 Summary and conclusions

Chapter 5 Approximate Bayesian Confidence Intervals For The Mean of a Gaussian Distribution corresponding to the Higgins-Tsokos loss function.

5.1 Introduction

5.2 Preliminaries

5.3 Main Results

5.4 Numerical results

5.5 Summary and conclusions

Chapter 6 Approximate Bayesian Confidence intervals for the Mean of a Gaussian Distribution Versus Bayesian models

6.1 Introduction

6.2 Preliminaries

6.3 Main Results

6.4 Numerical results

6.5 Summary and conclusions

Chapter 7 Approximate Bayesian Confidence Intervals for the Coefficient of Variation of a Gaussian distribution corresponding to The Square Error loss function.

7.1 Introduction

7.2 Preliminaries

7.3 Main Results

7.4 Numerical results

7.5 Summary and conclusions

Chapter 8 Approximate Bayesian Confidence Intervals for the Coefficient of Variation of a Gaussian distribution corresponding to The Higgins-Tsokos Loss function.

8.1 Introduction

8.2 Preliminaries

8.3 Main Results

8.4 Numerical results

8.5 Summary and conclusions

Chapter 9, Approximate Bayesian Confidence Intervals for The Mean of an Exponential Distribution corresponding to the Square Error loss function,

9.1 Introduction

9.2 Preliminaries

9.3 Main Results

9.4 Numerical results

9.5 Summary and conclusions

Chapter 10 Approximate Bayesian Confidence Intervals for the Mean of an Exponential Distribution corresponding to the Higgins-Tsokos loss function.

10.1 Introduction

10.2 Preliminaries

10.3 Main Results

10.4 Numerical results

10.5 Summary and conclusions

Chapter 11 Approximate Bayesian Confidence Intervals For The shape parameter of a Gamma Distribution.

11.1 Introduction

11.2 Preliminaries

11.3 Main Results

11.4 Numerical results

11.5 Summary and conclusions

Chapter 12 Approximate Bayesian Confidence Intervals For The scale and rate parameters of a Gamma Distribution

12.1 Introduction

12.2 Preliminaries

12.3 Main Results

12.4 Numerical results

12.5 Summary and conclusions

Preface

New And Efficient Approach and Closed-Form Confidence Intervals for Parameters of Normal, Exponential and Gamma Distributions provides a new approach and closed-form confidence bounds to estimating probability distribution parameters with great accuracy.

In this book I present new Approximate Bayesian confidence bounds for the parameters of Normal, Exponential and Gamma distributions.

Bayesian analysis implies the exploitation of suitable prior information and the choice of a loss function in association with Bayes' Theorem. It rests on the notion that a parameter within a model is not merely an unknown quantity but rather behaves as a 6random variable that follows some distribution.

Even though I found a beauty in the Bayesian theory, its sensitivity to the choice of the prior distribution was a great concern that led me to propose a new Approximate Bayesian approach which yields closed-form models with great coverage accuracy. Also, the new Approximate Bayesian approach and models rely only on the observations that are under study.

The goal of this book is to present a new and efficient parametric estimation approach to researchers, scientists, practitioners, engineers, educators and students.

Chapter 1 Introduction

The purpose of this book is to present a new Approximate Bayesian approach and new closed-form Approximate Bayesian confidence bounds for parameters of Normal, Exponential and Gamma distributions.

1.1 Bayesian Philosophy

Bayesian analysis implies the exploitation of suitable prior information and the choice of a loss function in association with Bayes' Theorem. It rests on the notion that a parameter within a model is not merely an unknown quantity but rather behaves as a random variable which follows some distribution. In the area of life testing, it is indeed realistic to assume that a life parameter is stochastically dynamic. This assertion is supported by the fact that the complexity of electronic and structural systems is likely to cause undetected component interactions resulting in an unpredictable fluctuation of life parameters. Recently, Drake (1966) gave an excellent account for the use of Bayesian statistics in reliability problems. As he points out " He (Bayesian) realizes that his selection of a prior (distribution) to express his present state of knowledge will necessarily be somewhat arbitrary. But he greatly appreciates this opportunity to make his entire assumptive structure clear to the world…" . "Why should an engineer not use his engineering judgment and prior knowledge about the parameters in the statistical distribution he has picked? For example, if it is the mean time between failures (MTBF) of an exponential distribution that must be evaluated from some tests, he undoubtedly has some idea of what the value will turn to be. If he does not, he is about to be fired. Then he can get much better idea about the true MTBF by combining some test results and his prior knowledge."

The loss functions that will be used are given below, along with a statement of their key characteristics.

Square Error loss function

The "popular" Square Error loss function places a small weight on estimates that are near the true value and proportionately more weight on extreme deviation from the true value of the parameter. Its popularity is due to its analytical tractability in Bayesian modeling. The Square Error loss is defined as follows:

$$L_{SE}(\hat{\theta},\theta) = \left(\hat{\theta}-\theta\right)^2$$

The corresponding Bayesian estimator of a parameter θ is defined as follows:

$$\hat{\theta} = \int \theta h(\theta/x)d\theta$$

where $h(\theta/X)$ is the posterior density function.

Higgins-Tsokos loss function

The Higgins-Tsokos loss function places a heavy penalty on extreme over- or underestimation. That is, it places an exponential weight on extreme errors. The Higgins-Tsokos loss function is defined as follows:

$$L_{HT}(\hat{\theta},\theta) = \frac{f_1 e^{f_2(\hat{\theta}-\theta)} + f_2 e^{-f_1(\hat{\theta}-\theta)}}{f_1+f_2} - 1, \; f_1, f_2 > 0.$$

The corresponding Bayesian estimator of a parameter θ is defined as follows:

$$\hat{\theta} = \frac{1}{f_1+f_2} Ln\left(\frac{\int e^{f_1\theta} h(\theta/x)d\theta}{\int e^{-f_2\theta} h(\theta/x)d\theta}\right)$$

where $h(\theta/X)$ is the posterior density function?.

1.2 Aims and Objectives

In this section, I will introduce the contents of the Chapters.

In Chapter 2, I will derive closed-form Approximate Bayesian confidence bounds for the variance of a Gaussian distribution, with the use of the Square Error loss function. The analytical development and numerical results show that the Approximate Bayesian approach and model rely only on the observations that are under study. The classical method that uses the Chi-square statistic does not always yield the best results. In fact, the Approximate Bayesian approach and confidence bounds perform often better than their classical counterparts.

In Chapter 3, I will derive closed-form Approximate Bayesian confidence bounds for the variance of a Gaussian distribution with the use of the Higgins-Tsokos loss function. The analytical development and numerical results show that the Approximate Bayesian approach and model rely only on the observations. The classical approach does not always yield the best confidence intervals. In fact, the Approximate Bayesian Approach and confidence intervals perform often better than their classical counterparts.

In Chapter 4 and Chapter 5, I will respectively derive closed-form Approximate Bayesian confidence bounds for the mean of a Gaussian distribution with the use of the Square Error and the Higgins-Tsokos loss functions. The analytical development and numerical results show that the Approximate Bayesian approach and confidence bounds rely only on the observations. The well–known classical method that uses the standard normal and the student-t statistics does not always yield the best results. In fact, the Approximate Bayesian approach and models perform often better.

Chapter 6 is concerned with the sensitivity of Bayesian analysis with respect to the selection of the prior distribution.

Using Normal data and SAS software, the Approximate Bayesian confidence intervals, for a Normal population mean, that were obtained in Chapter 4 and Chapter 5, will be compared to a published Bayesian model.

It is shown that The Bayesian model does not always yield the best confidence intervals. In fact, the Approximate Bayesian models perform better.

In Chapter 7 and Chapter 8, I will respectively derive closed-form Approximate Bayesian confidence bounds for the coefficient of variation of a Gaussian distribution, with the use of the Square Error and the Higgins-Tsokos loss functions. Using Normal data and SAS software, the Approximate Bayesian confidence intervals will be compared to a published classical model.

It is shown that the classical model does not always yield the best confidence intervals. In fact, the Approximate Bayesian confidence intervals have great coverage accuracy and perform often better.

In Chapter 9 and Chapter 10, I will respectively derive closed-form Approximate Bayesian confidence bounds for the mean of an Exponential distribution with the use of the Square Error and the Higgins-Tsokos loss functions. These models will be compared to the ones corresponding to a published classical model. The analytical development and numerical results show that the Approximate Bayesian approach and models rely only on the observations that are under study

. The classical model that uses the standard normal distribution does not always yield the best confidence intervals. In fact, the Approximate Bayesian confidence intervals have great coverage accuracy and perform often better.

In Chapter 11, I will derive closed-form Approximate Bayesian confidence bounds for the shape parameter of a two-parameter Gamma distribution, with the use of the Square Error loss function. The analytical developments and numerical results show that the Approximate Bayesian approach and confidence intervals have great coverage accuracy and rely only on the observations under study.

In Chapter 12, I will derive closed-form Approximate Bayesian confidence bounds for the scale and rate parameters of a two-parameter Gamma distribution, with the use of the Square Error.

The analytical developments and numerical results show that the Approximate Bayesian approach and models rely only on the observations and have great coverage accuracy.

Chapter 2 Approximate Bayesian Confidence Intervals for the Variance of a Gaussian Distribution corresponding to the Square Error Loss function

2.1 Introduction

The aim of this chapter is to derive closed-form confidence bounds for the variance of a Gaussian distribution, with the use of the Square Error loss function.

We shall consider a classical and useful underlying model. That is, we shall consider the Normal underlying model characterized by

$$f(x) = \frac{1}{\sqrt{2\pi}\sigma} e^{-\frac{1}{2}\left(\frac{x-\mu}{\sigma}\right)^2}; -\infty < x < \infty, -\infty < \mu < \infty, \sigma > 0.$$

Considering the Square Error loss function, we will derive Approximate Bayesian confidence intervals for the variance of a normal population.

To assess the performance of the obtained Approximate Bayesian confidence bounds, numerical results will be obtained with the use of Normal and approximately Normal data along with SAS software. The Approximate Bayesian results will then be compared to their classical counterparts corresponding to the well-known classical method.

Once the underlying model is found to be normally or approximately normally distributed, to construct confidence intervals for a Normal population variance, the well-known classical approach uses the following model that relies on the Chi-square statistic

$$\left(\frac{(n-1)s^2}{\chi^2_{n-1,\alpha/2}}, \frac{(n-1)s^2}{\chi^2_{n-1,1-\alpha/2}} \right)$$

:

We shall denote the inverse of the population variance σ^2 by θ and its corresponding estimate by $\hat{\theta}$.

2.2 Preliminaries

Although there is no specific analytical procedure that allows us to identify the appropriate loss function to be used, the most commonly used is the Square Error loss function. One of the reasons for selecting this loss function is because of its analytical tractability in Bayesian analysis.

In this Chapter we will consider the Square Loss function that is defined as follows:

$$L_{SE}(\hat{\theta},\theta)=\left(\hat{\theta}-\theta\right)^2$$

We will assume that θ behaves as a random variable which is characterized by the Pareto probability density function that is defined as follows:

$$f_1(\theta)=\frac{a}{b}\left(\frac{b}{\theta}\right)^{a+1} ; \theta \geq b \succ 0, a \succ 0.,$$

where $\theta=1/\sigma^2$.

The Pareto prior has been selected because of its mathematical tractability.

Let $x_1, x_2, \ldots, x_n$ denote the observations, of a given system, that are being characterized by the normal distribution.

Replacing $1/\sigma^2$ by θ, we obtain the following characterization of the Normal underlying probability distribution:.

$$f(x)=\frac{1}{\sqrt{2\pi\theta^{\frac{1}{2}}}}e^{-\theta\frac{(x-\mu)^2}{2}} ;-\infty \prec x \prec \infty, -\infty \prec \mu \prec \infty, \theta \succ 0$$

This leads to the following posterior distribution:

$$h(\theta \mid x) \frac{\theta^{\frac{n}{2}-a-1} e^{-\theta \frac{(x-\mu)^2}{2}}}{\int_b^\infty \theta^{\frac{n}{2}-a-1} e^{-\theta \frac{(x-\mu)^2}{2}} d\theta}, \theta \succ b..$$

2.3 Main results

2.3.1 Approximate Bayesian confidence bounds for σ^2 when the population mean is known.

With the use of sample data that are under study, the Pareto prior will be approximated in a manner that is suitable to yielding good Approximate Bayesian estimates of the parameter θ.

It is easily shown that the approximate Pareto prior

$$\bar{g}(\theta) = \frac{a1}{b1}\left(\frac{b1}{\theta}\right)^{a1+1}; a1 = \frac{n}{2} - 1, b1 = \frac{n-2}{\sum_{i=1}^n (x_i - \mu)^2}.$$

gives the following Approximate Bayesian estimator for the parameter θ:

$$\frac{n-1}{\sum_{i=1}^n (x_i - \mu)^2}$$

Using the corresponding approximate posterior distribution along with the equalities

$$P(\theta \succ L \mid x) = 1 - \alpha/2$$

and

$$P(\theta \succ U \mid x) = \alpha/2,$$

we obtain the following $100(1-\alpha)$ % lower and upper confidence bounds for θ:

$$L = \frac{n - 2 - 2Ln(1 - \alpha/2)}{\sum_{i=1}^{n}(x_i - \mu)^2} \qquad U = \frac{n - 2 - 2Ln(\alpha/2)}{\sum_{i=1}^{n}(x_i - \mu)^2},$$

Hence, the $100(1-\alpha)$% lower and upper Approximate Bayesian confidence bounds for the population variance σ^2 are the following:

$$L_{(SE)} = \frac{\sum_{i=1}^{n}(x_i - \mu)^2}{n - 2 - 2\ln(\alpha/2)}$$

$$U_{(SE)} = \frac{\sum_{i=1}^{n}(x_i - \mu)^2}{n - 2 - 2\ln(1 - \alpha/2)}$$

2.3.2 Approximate Bayesian confidence bounds for σ^2 when the population mean is unknown.

When the population mean is unknown, it is estimated by the sample mean $\bar{x}$.

Hence, when the population mean is not known, we have the following $100(1-\alpha)$ % lower and upper Approximate Bayesian confidence bounds for the population variance σ^2:

$$L_{b(SE)} = \frac{\sum_{i=1}^{n}(x_i - \bar{x})^2}{n - 2 - 2\ln(\alpha/2)}$$

$$U_{b(SE)} = \frac{\sum_{i=1}^{n}(x_i - \bar{x})^2}{n - 2 - 2\ln(1 - \alpha/2)}$$

2.4 Numerical results

To obtain numerical results, we will use samples that have been drawn from normally distributed populations (Examples 1, 2, 3, .4, 7) and approximately normal populations (Examples 5, 6). SAS software is also used to obtain the Normal population mean μ and standard deviation σ corresponding to each of the Normal and approximately Normal data sets that are given below. The lengths of the classical and Approximate Bayesian confidence intervals are respectively denoted by l_C and l_{SE}.

Example 1

Data obtained from Prem. S. Mann, Introductory Statistics, Third edition, page 504, 1998.

24, 28, 22, 25, 24, 22, 29, 26, 25, 28, 19, 29.

Normal population distribution obtained with SAS:

$$N(\mu = 25.083, \sigma = 3.1176)$$

Population and sample variances:

$$\sigma^2 = 9.71943, \quad s^2 = 9.719696$$

Table 4.1: Classical and Approximate Bayesian confidence intervals for σ^2 corresponding to the first data set.

Confidence level	Classical bounds	Approx. Bayesian bounds (SE)

80%	6.1890 – 19.1675	7.3204 – 10.4710
90%	5.4341 – 23.3697	6.6858 – 10.5830
95%	4.8775 – 28.0171	6.1525 – 10.6378
99%	3.9958 – 41.0744	5.1909 – 10.6809

Confidence level	$(l_C) \div (l_{SE})$
80%	4.1193
90%	4.6021
95%	5.1589
99%	6.7538

Example 2

Data obtained from Prem. S. Mann, Introductory Statistics, Third edition, page 504, 1998.

13, 11, 9, 12, 8, 10, 5, 10, 9, 12, 13.

Normal population distribution obtained with SAS :

$N(\mu = 10.182, \sigma = 2.4008)$

Population and sample variances:

$\sigma^2 = 5.76384$, $s^2 = 5.763636$

Table 4.2: Classical and Approximate Bayesian confidence intervals for σ^2 corresponding to the second data set.

Confidence level	Classical bounds	Approx. Bayesian bounds (SE)
80%	3.6052 – 11.8472	4.2363 – 6.2575
90%	3.1483 – 14.6285	3.8446 – 6.3318
95%	2.8139 – 17.7506	3.5191 – 6.3682
99%	2..2885 – 26.7330	2.9411 – 6.3969

Confidence level	$(l_C) \div (l_{SE})$
80%	4.0777
90%	4.6157
95%	5.2426
99%	7.0734

Example 3

Data obtained from Prem. S. Mann, Introductory Statistics, Third edition, page 504, 1998.

16, 14, 11, 19, 14, 17, 13, 16, 17, 18, 19, 12.

Normal population distribution obtained with SAS:

$N(\mu = 15.5, \sigma = 2.6799)$

Population and sample variances:

$\sigma^2 = 7.18186$, $s^2 = 7.181818$

Table 4.3: Classical and Approximate Bayesian confidence intervals for σ^2 corresponding to the third data set.

Confidence level	Classical bounds	Approx. Bayesian bounds (SE)
80%	4.5731 – 14.1627	5.4090 – 7.7369
90%	4.0152 – 17.2677	4.9401 – 7.8197
95%	3.6040 – 20.7023	4.5460 – 7.8601
99%	2.9524 – 30.3496	3.8355 – 7.8920

Confidence level	$(l_C) \div (l_{SE})$
80%	4.1194
90%	4.6022
95%	5.1592
99%	6.7539

Example 4

Data obtained from Prem. S. Mann, Introductory Statistics, Third edition, page 504, 1998.

27, 31, 25, 33, 21, 35, 30, 26, 25, 31, 33, 30, 28.

Normal population distribution obtained with SAS:

$N(\mu = 28.846, \sigma = 3.9549)$

Population and sample variances:

$\sigma^2 = 15.64123$, $s^2 = 15.641025$

Table 4.4: Classical and Approximate Bayesian confidence intervals for σ^2 corresponding to the fourth data set.

Confidence level	Classical bounds	Approx.Bayesian bounds (SE)
80%	10.1187 – 29.7735	12.0275 – 16.7422
90%	8.9266 – 35.9151	11.0462 – 16.9052
95%	8.0426 – 42.6186	10.2130 – 16.9847
99%	6.6322 – 61.0580	8.6908 – 17.0474

Confidence level	$(l_C) \div (l_{SE})$
80%	*4.1688*
90%	*4.6063*
95%	*5.1059*
99%	*6.5129*

Example 5

Data obtained from James T. McClave/Terry Sincich, A first course in Statistics, page 301, Sixth edition, 1997.

52, 33, 42, 44, 41, 50, 44, 51, 45, 38, 37, 40, 44, 50, 43.

Normal population distribution obtained with SAS:

$N(\mu = 43.6, \sigma = 5.4746)$

Population and sample variances:

$\sigma^2 = 29.97124$, $s^2 = 29.971428$

Table 4.5: Classical and Approximate Bayesian confidence intervals for σ^2 corresponding to the fifth data set.

Confidence level	Classical bounds	Approx.Bayesian bounds (SE)
80%	19.9202 – 53.8639	23.8339 – 31.7620
90%	17.7159 – 63.8563	22.0941 – 32.0242
95%	16.0650 – 74.5426	20.5910 – 32.1516
99%	13.3976 – 102.9693	17.7821 – 32.2520

Confidence level	$(l_C) \div (l_{SE})$
80%	4.2814
90%	4.6465
95%	5.0583
99%	6.1902

Example 6

Data obtained from James T. McClave/Terry Sincich, A first course in Statistics, page 301, Sixth edition, 1997.

52, 43, 47, 56, 62, 53, 61, 50, 56, 52, 53, 60, 50, 48, 60, 55.

Normal population distribution obtained with SAS:

$N(\mu = 53.625, \sigma = 5.4145)$

Population and sample variances:

$\sigma^2 = 29.31681$, $s^2 = 29.316666$

Table 4.6: Classical and Approximate Bayesian confidence intervals for σ^2 corresponding to the sixth data set.

Confidence level	Classical bounds	Approx.Bayesian bounds (SE)
80%	19.7135 - 51.4508	23.6359 – 30.9449
90%	17.5928 – 60.5632	21.9968 – 31.1822
95%	15.9978 – 70.2252	20.5794 – 31.2975
99%	13.4066 – 95.5771	17.8784 – 31.3882

Confidence level	$(l_C) \div (l_{SE})$
80%	4.3422
90%	4.6781
95%	5.0551
99%	6.0822

Example 7

The following observations have been obtained from the collection of SAS data sets.

50, 65, 100, 45, 111, 32, 45, 28, 60, 66, 114, 134, 150, 120, 77, 108, 112, 113, 80, 77, 69, 91, 116, 122, 37, 51, 53, 131, 49, 69, 66, 46, 131, 103, 84, 78.

Normal population distribution obtained with SAS:

$N(\mu = 82.861, \sigma = 33.226)$

Population and sample variances:

$\sigma^2 = 1103.96716$, $s^2 = 1103.951587$.

Table 4.7: Classical and Approximate Bayesian confidence intervals for σ^2 corresponding to the seventh data set.

Confidence level	Classical bounds	Approx.Bayesian bounds (SE)
80%	839.407 – 1556.427	1000.85 – 1129.42
90%	776.408 – 1717.182	966.16 – 1133.00
95%	726.824 – 1874.554	933.79 – 1134.73
99%	641.630 – 2240.291	866.39 – 1136.08

Confidence level	$(l_C) \div (l_{SE})$
80%	5.5772
90%	5.6388
95%	5.7119
99%	5.9277

All the above tables show that the obtained Approximate Bayesian confidence intervals contain the population variance σ^2 and are strictly included in their classical counterparts; also, the lengths of the classical confidence intervals are more than four times greater than their counterparts corresponding to the new Approximate Bayesian approach.

2.5 Summary and Conclusion

i) The classical method used to construct confidence intervals for the variance of a Normal population does not always yield the best coverage accuracy. In fact, the above Approximate Bayesian confidence intervals contain the population variance, and are all strictly included in their classical counterparts.

ii) Contrary to the classical method that uses the Chi-square statistic, the new Approximate Bayesian approach and confidence bounds rely only on the observations that are under study...

iii) With the new Approximate Bayesian approach, Approximate Bayesian confidence intervals for a Normal or approximately Normal population variance are easily constructed for any level of significance.

Chapter 3 Approximate Bayesian Confidence Intervals for the Variance of a Gaussian distribution corresponding to The Higgins-Tsokos loss function

3.1   Introduction

The aim of this chapter is to construct closed-form Approximate Bayesian confidence bounds for the variance of a Gaussian distribution, with the use of Higgins-Tsokos Loss function.

Once the underlying model has been selected along with a prior, most Bayesian results are obtained with the use of the Square Error loss function. Although there is no specific analytical procedure that allows us to identify the appropriate loss function to employ, the most commonly used is the Square Error loss.

In this chapter, the Higgins-Tsokos loss function will be used.

Considering the normal underlying model

$$f(x) = \frac{1}{\sqrt{2\pi}\sigma} e^{-\frac{1}{2}\left(\frac{x-\mu}{\sigma}\right)^2} ; -\infty < x < \infty, -\infty < \mu < \infty, \sigma > 0.$$

and the Higgins-Tsokos loss function, we will derive closed-from Approximate Bayesian confidence bounds for the variance of a normal population σ^2.

To assess the performance of the obtained Approximate Bayesian confidence intervals, numerical results will be obtained with the use of Normal and approximately Normal data along with SAS software. The Approximate Bayesian results will then be compared to their classical counterparts corresponding to the well-known classical method.

As mentioned in Chapter 3, once the underlying model is found to be normally or approximately normally distributed, to construct confidence intervals for a Normal population variance, the well-known classical approach uses the following model:

$$\left(\frac{(n-1)s^2}{\chi^2_{n-1,\alpha/2}}, \frac{(n-1)s^2}{\chi^2_{n-1,1-\alpha/2}} \right)$$

We shall denote the inverse of the population variance σ^2 by θ and its corresponding estimate by $\hat{\theta}$.

3.2 Preliminaries

The Higgins-Tsokos loss function is defined as follows:

$$L_{HT}(\hat{\theta},\theta) = \frac{f_1 e^{f_2(\hat{\theta}-\theta)} + f_2 e^{-f_1(\hat{\theta}-\theta)}}{f_1 + f_2} - 1, f_1, f_2 \succ 0.$$

We will assume that θ behaves as a random variable that is characterized by the Pareto probability density function which is defined as follows:

$$f_1(\theta) = \frac{a}{b}\left(\frac{b}{\theta}\right)^{a+1}; \theta \geq b \succ 0, a \succ 0. \text{ , where } \theta = 1/\sigma^2.$$

Let $x_1, x_2, \ldots, x_n$ denote the observations of a given system that are being characterized by the normal distribution.

Replacing $1/\sigma^2$ by θ, we obtain the following characterization of the Normal underlying probability distribution:.

$$f(x) = \frac{1}{\sqrt{2\pi}\theta^{\frac{1}{2}}} e^{-\theta\frac{(x-\mu)^2}{2}}; -\infty < x < \infty, -\infty < \mu < \infty, \theta > 0$$

This leads to the following posterior distribution:

$$h(\theta \mid x) \frac{\theta^{\frac{n}{2}-a-1} e^{-\theta\frac{(x-\mu)^2}{2}}}{\int_b^\infty \theta^{\frac{n}{2}-a-1} e^{-\theta\frac{(x-\mu)^2}{2}} d\theta}, \theta > b..$$

3.3 Main results

3.3.1 Approximate Bayesian confidence bounds for σ^2 when the population mean μ is known.

With the use of sample data that are under study, the Pareto prior will be approximated in a manner that is suitable to yielding good Approximate Bayesian estimates of the parameter θ.

It is easily shown that the approximate Pareto prior

$$\bar{g}_1(\theta) = \frac{a_0}{b_0} \left(\frac{b_0}{\theta}\right)^{a_0+1}; a_0 = \frac{n}{2} - 1, b_0 = \frac{n-1}{\sum_{i=1}^n (x_i - \mu)^2} - \frac{1}{f_1 + f_2} Ln\left(\frac{\sum_{i=1}^n \frac{(x-\mu)^2}{2} + f_2}{\sum_{i=1}^n \frac{(x-\mu)^2}{2} - f_1}\right),$$

$$f_1 < \frac{(x-\mu)^2}{2}$$

yields the following Approximate Bayesian estimate of parameter θ:

$$\frac{n-1}{\sum_{i=1}^{n}(x_i - \mu)^2}$$

Using the corresponding approximate posterior distribution along with the equalities

$$P(\theta \succ L| x) = 1 - \alpha/2$$

and

$$P(\theta \succ U| x) = \alpha/2,$$

we obtain the following $100(1-\alpha)$ % lower and upper Approximate Bayesian confidence bounds for θ:

$$L = \frac{n-1-2Ln(1-\alpha/2)}{\sum_{i=1}^{n}(x_i - \mu)^2} - \frac{1}{f_1 + f_2} Ln\left(\frac{\sum_{i=1}^{n}(x_i - \mu)^2 + f_2}{\sum_{i=1}^{n}(x_i - \mu)^2 - f_1}\right)$$

$$U = \frac{n-1-2Ln(\alpha/2)}{\sum_{i=1}^{n}(x_i - \mu)^2} - \frac{1}{f_1 + f_2} Ln\left(\frac{\sum_{i=1}^{n}(x_i - \mu)^2 + f_2}{\sum_{i=1}^{n}(x_i - \mu)^2 - f_1}\right).$$

Thus, we have the following $100(1-\alpha)$ % Approximate Bayesian confidence bounds for the normal population variance, when the population mean is known.

$$L_{b(HT)} = \cfrac{1}{\cfrac{n-1-2Ln(\alpha/2)}{\sum_{i=1}^{n}(x_i - \mu)^2} - \cfrac{1}{f_1 + f_2} Ln\left(\cfrac{\sum_{i=1}^{n}(x_i - \mu)^2 + f_2}{\sum_{i=1}^{n}(x_i - \mu)^2 - f_1}\right)}$$

$$U_{b(HT)} = \cfrac{1}{\cfrac{n-1-2Ln(1-\alpha/2)}{\sum_{i=1}^{n}(x_i - \mu)^2} - \cfrac{1}{f_1 + f_2} Ln\left(\cfrac{\sum_{i=1}^{n}(x_i - \mu)^2 + f_2}{\sum_{i=1}^{n}(x_i - \mu)^2 - f_1}\right)}.$$

3.3.2 Approximate Bayesian confidence bounds of σ^2 when the population mean μ is unknown.

When the population mean is unknown, it is estimated by the sample mean $\bar{x}$.

Hence, when the population mean is not known, we have the following $100(1-\alpha)\%$ lower and upper Approximate Bayesian confidence bounds for the population variance σ^2:

$$L_{b(HT)} = \cfrac{1}{\cfrac{n-1-2Ln(\alpha/2)}{\sum_{i=1}^{n}(x_i - \bar{x})^2} - \cfrac{1}{f_1 + f_2} Ln\left(\cfrac{\sum_{i=1}^{n}(x_i - \bar{x})^2 + f_2}{\sum_{i=1}^{n}(x_i - \bar{x})^2 - f_1}\right)}$$

$$U_{b_{(HT)}} = \cfrac{1}{\cfrac{n-1-2Ln(1-\alpha/2)}{\sum_{i=1}^{n}(x_i-\bar{x})^2} - \cfrac{1}{f_1+f_2} Ln\left(\cfrac{\sum_{i=1}^{n}(x_i-\bar{x})^2 + f_2}{\sum_{i=1}^{n}(x_i-\bar{x})^2 - f_1}\right)}.$$

3.4 Numerical results

For the numerical results, we will use samples that have been drawn from normally distributed populations (Examples 1, 2, 3, .4, 7) and approximately normal populations (Examples 5, 6). SAS software is used to obtain the normal population mean μ and standard deviation σ corresponding to each of the Normal and approximately Normal data sets that are given below. The lengths of the classical and Approximate Bayesian confidence intervals are respectively denoted by l_C and l_{HT}.

Example 1

Data obtained from Prem. S. Mann, Introductory Statistics, Third edition, page 504, 1998.

24, 28, 22, 25, 24, 22, 29, 26, 25, 28, 19, 29.

Normal population distribution obtained with SAS:

$N(\mu = 25.083, \sigma = 3.1176)$

Population and sample variances:

$\sigma^2 = 9.71943$, $s^2 = 9.719696$

Table 4.1: Classical and Approximate Bayesian confidence intervals for σ^2 corresponding to the first data set.

Confidence level	Classical bounds	Approx. Bayesian bounds (HT)
80%	6.1890 – 19.1675	8.0881– 10.2349

90%	5.4341 – 23.3697	7.3204 –10.4710
95%	4.8775 – 28.0171	6.6858 –10.5831
99%	3.9958– 41.0744	5.5655 –10.6702

Confidence level	$(l_C) \div (l_{HT})$
80%	6.0455
90%	5.6927
95%	5.9373
99%	7.2636

Example 2

Data obtained from Prem. S. Mann, Introductory Statistics, Third edition, page 504, 1998.

13, 11, 9, 12, 8, 10, 5, 10, 9, 12, 13.

Normal population distribution obtained with SAS :

$N(\mu = 10.182, \sigma = 2.4008)$

Population and sample variances:

$\sigma^2 = 5.76384$, $s^2 = 5.763636$

Table 4.2: Classical and Approximate Bayesian confidence intervals for σ^2 corresponding to the second data set.

Confidence level	Classical bounds	Approx. Bayesian bounds (HT)
80%	3.6052 – 11.8472	4.7170 – 6.1015
90%	3.1483 – 14.6285	4.2363 – 6.2575
95%	2.8139 – 17.7506	3.8446 – 6.3319
99%	2.2885 – 26.7330	3.1650 – 6.3898

Confidence level	$(l_C) \div (l_{HT})$
80%	5.9530
90%	5.6804
95%	6.0051
99%	7.5801

Example 3

Data obtained from Prem. S. Mann, Introductory Statistics, Third edition, page 504, 1998.

16, 14, 11, 19, 14, 17, 13, 16, 17, 18, 19, 12.

Normal population distribution obtained with SAS:

$$N(\mu = 15.5, \sigma = 2.6799)$$

Population and sample variances:

$\sigma^2 = 7.18186$, $s^2 = 7.181818$

Table 4.3: Classical and Approximate Bayesian confidence intervals f0r σ^2 corresponding to the third data set.

Confidence level	Classical bounds	Approx.Bayesian bounds (HT)
80%	4.5731 – 14.1627	5.9763 – 7.5625
90%	4.0152 – 17.2677	5.4090 – 7.7370
95%	3.6040 – 20.7023	4.9401 – 7.8198
99%	2.9524 – 30.3496	4.1123 – 7.8841

Confidence level	$(l_C) \div (l_{HT})$
80%	6.0456
90%	5.6926
95%	5.9375
99%	7.2636

Example 4

Data obtained from Prem. S. Mann, Introductory Statistics, Third edition, page 504, 1998.

27, 31, 25, 33, 21, 35, 30, 26, 25,31.33.30, 28.

Normal population distribution obtained with SAS:

$N(\mu = 28.846, \sigma = 3.9549)$

Population and sample variances:

$\sigma^2 = 15.64123$, $s^2 = 15.641025$

Table 4.4: Classical and Approximate Bayesian confidence intervals for σ^2 corresponding to the fourth data set.

Confidence level	Classical bounds	Approx. Bayesian bounds (HT)
80%	10.1187 – 29.7735	13.2002 – 16.3976
90%	8.9266 – 35.9151	12.0275 – 16.7422
95%	8.0426 – 42.6186	11.0462 – 16.9052
99%	6.6322 – 61.0580	9.2869 – 17.0318

Confidence level	$(l_C) \div (l_{HT})$
80%	6.1471
90%	5.7243
95%	5.9013
99%	7.0273

Example 5

Data obtained from James T. McClave/Terry Sincich, A first course in Statistics, page 301, Sixth edition, 1997.

52, 33, 42, 44, 41, 50, 44, 51, 45, 38, 37, 40, 44, 50, 43.

Normal population distribution obtained with SAS:

$N(\mu = 43.6, \sigma = 5.4746)$

Population and sample variances:

$\sigma^2 = 29.97124$, $s^2 = 29.971428$

Table 4.5: Classical and Approximate Bayesian confidence intervals for σ^2 corresponding to the fifth data set.

Confidence level	Classical bounds	Approx.Bayesian bounds (HT)
80%	19.9202 – 53.8639	25.8710 – 31.2056
90%	17.7159 – 63.8563	23.8339 – 31.7620
95%	16.0650 – 74.5426	22.0941 – 32.0242
99%	13.3976 – 102.9693	18.8921 – 32.2270

Confidence level	$(l_C) \div (l_{HT})$
80%	6.3629
90%	5.8198
95%	5.8889
99%	6.7170

Example 6

Data obtained from James T. McClave/Terry Sincich, A first course in Statistics, page 301, Sixth edition, 1997.

52, 43, 47, 56, 62, 53, 61, 50, 56, 52, 53, 60, 50, 48, 60, 55.

Normal population distribution obtained with SAS:

$N(\mu = 53.625, \sigma = 5.4145)$

Population and sample variances:

$\sigma^2 = 29.31681$, $s^2 = 29.316666$

Table 4.6: Classical and Approximate Bayesian confidence intervals for σ^2 corresponding to the sixth data set.

Confidence level	Classical bounds	Approx.Bayesian bounds (HT)
80%	19.7135 - 51.4508	25.5388 – 30.4403
90%	17.5928 – 60.5632	23.6359 – 30.9495
95%	15.9978 – 70.2252	21.9968 – 31.1822
99%	13.4066 – 95.5771	18.9462 – 31.3656

Confidence level	$(l_C) \div (l_{HT})$
80%	6.4743
90%	5.8754
95%	5.9036
99%	6.6163

Example 7

The following observations have been obtained from the collection of SAS data sets.

50, 65, 100, 45, 111, 32, 45, 28, 60, 66, 114, 134, 150, 120, 77, 108, 112, 113, 80, 77, 69, 91, 116, 122, 37, 51, 53, 131, 49, 69, 66, 46, 131, 103, 84, 78.

Normal population distribution obtained with

SAS: $N(\mu = 82.861, \sigma = 33.226)$

Population and sample variances:

$\sigma^2 = 1103.96716$, $s^2 = 1103.951587$.

Table 4.7: Classical and Approximate Bayesian confidence intervals for σ^2 corresponding to the seventh data set.

Confidence level	Classical bounds	Approx. Bayesian bounds (HT)
80%	839.407 – 1556.427	1038.13 – 1121.69
90%	776.408 – 1717.182	1000.85 – 1129.42
95%	726.824 – 1874.554	966.16 – 1133.00
99%	641.630 – 2240.291	894.19 – 1135.74

Confidence level	$(l_C) \div (l_{HT})$
80%	8.5808
90%	7.3176
95%	6.8792
99%	6.6181

All the above tables show that the obtained Approximate Bayesian confidence intervals contain the population variance σ^2 and are strictly included in their classical counterparts; also, the lengths of the classical confidence intervals are more than five times greater than the ones corresponding to the new Approximate Bayesian approach.

3.5 Summary and Conclusion

iv) The classical method used to construct confidence intervals for the variance of a Normal population does not always yield the best coverage accuracy. In fact, the obtained Approximate Bayesian confidence intervals contain the population variance and are all strictly included in their classical counterparts.

v) Contrary to the classical method that uses the Chi-square statistic, the new Approximate Bayesian approach and confidence bounds rely only on the observations that are under study...

vi) With the new Approximate Bayesian approach, Approximate Bayesian confidence intervals for a normal population variance are easily obtained for any level of significance.

Chapter 4   Approximate Bayesian Confidence Intervals For The Mean of a Gaussian Distribution corresponding to the Square Error Loss

4.1 Introduction

The aim of Chapter 4 is to construct closed-form Approximate Bayesian confidence intervals for the mean of a Gaussian distribution.

Considering the normal underlying model

$$f(x) = \frac{1}{\sqrt{2\pi}\sigma} e^{-\frac{1}{2}\left(\frac{x-\mu}{\sigma}\right)^2}; -\infty \prec x \prec \infty, -\infty \prec \mu \prec \infty, \sigma \succ 0.$$

and the Square Error loss function, we will derive Approximate Bayesian confidence bounds for the mean of a normal population.

Once the underlying model is found to be normally or approximately normally distributed, to construct confidence intervals for a Normal population mean, the well-known classical approach uses the following models that rely on the standard Normal and the student-t statistics:

$$\left(\overline{X} - t_{n-1,\,\alpha/2} \frac{s}{\sqrt{n}},\, \overline{X} + t_{n-1,\,\alpha/2} \frac{s}{\sqrt{n}} \right)$$

$$\left(\overline{X} - Z_{\alpha/2} \frac{\sigma}{\sqrt{n}},\, \overline{X} + Z_{\alpha/2} \frac{\sigma}{\sqrt{n}} \right)$$

4.2 Preliminaries

Considering the above Normal density function, to derive our approximate Bayesian confidence bounds for the mean of a normal distribution, we will use the following results that we obtained in Chapter 2:

Confidence bounds for a normal population variance

$$L_{\sigma^2(SE)} = \frac{\sum_{i=1}^{n}(x_i - \bar{x})^2}{n - 2 - 2\ln(\alpha/2)}$$

$$U_{\sigma^2(SE)} = \frac{\sum_{i=1}^{n}(x_i - \bar{x})^2}{n - 2 - 2\ln(1 - \alpha/2)}$$

4.3 Main Results

Using the above Approximate Bayesian confidence bounds for a normal population along with

$$\sigma^2 = E(X^2) - \mu^2$$

we can easily obtain the following Approximate Bayesian confidence bounds for a strictly positive mean of a normal population: variance along with:

$$U_{\mu(SE)} = \left(\frac{\sum_{i=1}^{n}(x_i - \bar{x})^2}{n-1} + \bar{x}^2 - \frac{\sum_{i=1}^{n}(x_i - \bar{x})^2}{n-2-2\ln(\alpha/2)} \right)^{0.5}$$

$$L_{\mu(SE)} = \left(\frac{\sum_{i=1}^{n}(x_i - \bar{x})^2}{n-1} + \bar{x}^2 - \frac{\sum_{i=1}^{n}(x_i - \bar{x})^2}{n-2-2\ln(1-\alpha/2)} \right)^{0.5}$$

Hence, for a normal random variable X with a mean that is smaller or equal to zero, we can infer the following Approximate Bayesian confidence bounds:

$$U_{\mu(SE)} = \left(\frac{\sum_{i=1}^{n}(y_i - \bar{y})^2}{n-1} + \bar{y}^2 - \frac{\sum_{i=1}^{n}(y_i - \bar{y})^2}{n-2-2\ln(\alpha/2)} \right)^{0.5} - a$$

$$L_{\mu(SE)} = \left(\frac{\sum_{i=1}^{n}(y_i - \bar{y})^2}{n-1} + \bar{y}^2 - \frac{\sum_{i=1}^{n}(y_i - \bar{y})^2}{n-2-2\ln(1-\alpha/2)} \right)^{0.5} - a$$

where y=x+a and "a" is a constant such that x+a>0

4.4 Numerical Examples and Results

For the numerical results, we will use samples that have been obtained from normally distributed populations (Examples 1, 2, 3, .4, 7) and approximately normal populations (Examples 5, 6) .SAS software is used to obtain the normal population mean μ and standard deviation σ corresponding to each of the Normal and approximately Normal data sets that are given below. The lengths of the classical and Approximate Bayesian confidence intervals are respectively denoted by l_C and l_{SE} .

Example 1

Data obtained from Prem. S. Mann, Introductory Statistics, Third edition, page 504, 1998

24, 28, 22, 25, 24, 22, 29, 26, 25, 28, 19, 29.

Normal population distribution obtained with SAS:

$N(\mu = 25.083, \sigma = 3.1176)$

$s^2 = 9.719696$

The corresponding sample mean and sample variance are

$\bar{x} = 25.08333$

Table 1: Classical and Approximate Bayesian confidence intervals for the population mean corresponding to the first example of data set.

C. L. %	Approx.Bayesian bounds (SE)	Classical bounds	WC WSE
80	25.0683-25.1311	23.85665-26.31001	39.87
90	25.0661-25.1437	23.46696-26.69971	41.66
95	25.0650-25.1543	23.10246-27.06420	44.36
99	25.0641-25.1734	22.28798-27.87869	51.15

Example 2

Data obtained from Prem. S. Mann, Introductory Statistics, Third edition, page 504, 1998

13, 11, 9, 12, 8, 10, 5, 10, 9, 12, 13.

Normal population distribution obtained with SAS:

$$N(\mu = 10.182, \sigma = 2.4008)$$

The corresponding sample mean and sample variance are

$$\bar{x} = 10.181812$$

$$s^2 = 5.763636$$

Table 2 Classical and Approximate Bayesian confidence intervals for the population mean corresponding to the second example of data set.

C.L. %	Approximate Bayesian bounds (SE)	Classical bounds	WC WSE
0	10.1575-10.2565	9.18869-11.17495	20.06
90	10.1538-10.2756	8.87019-11.49344	21.54
95	10.1520-10.2914	8.56907-11.79457	23.14
99	10.1506-10.3194	7.88792-12.47572	27.18

Example 3

Data obtained from Prem. S. Mann, Introductory Statistics, Third edition, page 504, 1998.

16, 14, 11, 19, 14, 17, 13, 16, 17, 18, 19, 12.

$N(\mu = 15.5, \sigma = 2.6799)$
Normal population distribution obtained with SAS:

The corresponding sample mean and sample variance are

$\bar{x} = 15.5$

$s^2 = 7.181818$

Table 3: Classical and Approximate Bayesian confidence intervals for the population mean corresponding to the third example of data set.

C.L. %	Approx. Bayesian bounds (SE)	Classical bounds	WC WSE
80	15.4820-15.5570	14.44556-16.55440	28.12
90	15.4794-15.5721	14.11058-16.88942	29.98
95	15.4781-15.5847	13.79727-17.20273	31.95
99	15.4770-15.6075	13.09714-17.90286	36.83

Example 4

Data obtained from Prem. S. Mann, Introductory Statistics, Third edition, page 504, 1998.

27, 31, 25, 33, 21, 35, 30, 26, 25, 31. 33. 30, 28.

Normal population distribution obtained with SAS:

$$N(\mu = 28.846, \sigma = 3.9549)$$

The corresponding sample mean and sample variance are

$$s^2 = 15.641025$$

$$\bar{x} = 28.846153$$

Table 4: Classical and Approximate Bayesian confidence intervals for the population mean corresponding to the fourth example of data set.

C.L. %	Approximate Bayesian bounds (SE)	Classical bounds	WC WSE
80	28.8270-28.9087	27.35878-30.33353	36.41
90	28.8242-28.9256	26.89151-30.80080	38.55
95	28.8228-28.9400	26.45604-31.2362	40.79
99	28.8217-28.9663	25.49517-32.19714	46.35

Example 5

Data obtained from James T. McClave/Terry Sincich, A first course in Statistics, page 301, Sixth edition, 1997

52, 33, 42, 44, 41, 50, 44, 51, 45, 38, 37, 40, 44, 50, 43.

$N(\mu = 43.6, \sigma = 5.4746)$
Normal population distribution obtained with SAS:

$\bar{x} = 43.6$
The corresponding sample mean and sample variance are

$s^2 = 29.971428$

Table 5: Classical and Approximate Bayesian confidence intervals for the population mean corresponding to the fifth example of data set.

C.L. %	Approximate Bayesian bounds (SE)	Classical bounds	WC WSE
80	43.5794-43.6703	41.69879-45.50121	41.83
90	43.5764-43.6902	41.11076-46,08924	43.75
95	43.5749-43.7074	40.56796-46.63204	63.30
99	43.5738-43.7395	39.39189-47.80811	50.79

Example 6

Data obtained from James T. McClave/Terry Sincich, A first course in Statistics, page 301, Sixth edition, 1997

52, 43, 47, 56, 62, 53, 61, 50, 56, 52, 53, 60, 50, 48, 60, 55.

Normal population distribution obtained with SAS:

$N(\mu = 53.625, \sigma = 5.4145)$

$\bar{x} = 53.625$
The corresponding sample mean and sample variance are

$s^2 = 29.316666$

Table 6: Classical and Approximate Bayesian confidence intervals for the population mean corresponding to the sixth example of data set.

C. L. %	Approximate Bayesian bounds (SE)	Classical bounds	WC WSE
80	53.6098-53.6779	51.80979-55.44021	53.31
90	53.6076-53.6932	51.25210-55.99790	55.44
95	53.6065-53.7064	50.74043-56.50957	57.75
99	53.6056-53.7315	49.63588-57.61412	63.37

Example 7

The following observations have been obtained from the collection of SAS data sets.

50, 65, 100, 45, 111, 32, 45, 28, 60, 66, 114, 134, 150, 120, 77, 108, 112, 113, 80, 77, 69, 91, 116, 122, 37, 51, 53, 131, 49, 69, 66, 46, 131, 103, 84, 78.

Normal population distribution obtained with S

$N(\mu=82.861, \sigma=33.226)$

The corresponding sample mean and sample variance are

$\bar{x} = 82.8611$

$s^2 = 1103.951587$

Table 7: Classical and Approximate Bayesian confidence intervals for the population mean corresponding to the seventh example of data set.

C.L. %	Approximate Bayesian bounds (SE)	Classical bounds	WC WSE
80	82.7072-83.4808	75.6261-90.0959	18.70
90	82.6856-83.6884	73.5052-92.2168	18.66
95	82.6751-83.8815	71.6196-94.1024	18.64
99	82.6669-84.2823	67.7793-97.9427	18.67

All the above tables show that the obtained Approximate Bayesian confidence intervals contain the population mean and are strictly included in their classical counterparts; also, the widths of the classical confidence intervals are more than twenty times greater than the ones corresponding to their Approximate Bayesian counterparts.

4.5 Summary and Conclusion

i) The classical and the Approximate Bayesian models perform well.

ii) The classical method used to constructing confidence intervals for the mean of a Normal population does not always yield the best coverage accuracy. In fact, the above Approximate Bayesian confidence intervals perform better than their classical counterparts.

iii)   Contrary to the classical method that uses the standard Normal and the student-t statistics, the new Approximate Bayesian approach and confidence bounds rely only on the observations that are under study...

iv)   With the new Approximate Bayesian approach, Approximate Bayesian confidence intervals for a Normal or approximately Normal population mean are easily obtained for any level of significance.

Chapter 5 Approximate Bayesian Confidence Intervals For The Mean of a Gaussian Distribution Corresponding to the Higgins-Tsokos loss function

5.1 Introduction

In Chapter 5, we will construct closed-form Approximate Bayesian confidence bounds for the mean of a Gaussian distribution, with the use of the Higgins-Tsokos loss function.

Considering the normal underlying model

$$f(x) = \frac{1}{\sqrt{2\pi}\sigma} e^{-\frac{1}{2}\left(\frac{x-\mu}{\sigma}\right)^2}; -\infty < x < \infty, -\infty < \mu < \infty, \sigma > 0$$

and the Higgins-Tsokos loss function, we will derive Approximate Bayesian confidence bounds for the mean of a Gaussian distribution.

To assess the performance of the new Approximate Bayesian model, numerical results will be obtained with the use of Normal and approximately Normal data along with SAS software. The Approximate Bayesian results will then be compared to their classical counterparts corresponding to the following models:

$$\left(\bar{X} - Z_{\alpha/2} \frac{\sigma}{\sqrt{n}}, \bar{X} - Z_{\alpha/2} \frac{\sigma}{\sqrt{n}}\right)$$

$$\left(\bar{X} - t_{n-1,\,\alpha/2} \frac{s}{\sqrt{n}}, \bar{X} + t_{n-1,\,\alpha/2} \frac{s}{\sqrt{n}}\right)$$

5.2 Preliminaries

Considering the above Normal density function, to derive the Approximate Bayesian confidence bounds for the mean of a Normal distribution, we will use results that were obtained in Chapter 3; we will use the following Approximate Bayesian confidence bounds for a normal population variance:

$$L_{b(HT)} = \frac{1}{\dfrac{n-1-2Ln(\alpha/2)}{\sum_{i=1}^{n}(x_i-\bar{x})^2} - \dfrac{1}{f_1+f_2} Ln\left(\dfrac{\sum_{i=1}^{n}(x_i-\bar{x})^2 + f_2}{\sum_{i=1}^{n}(x_i-\bar{x})^2 - f_1}\right)}$$

$$U_{b(HT)} = \frac{1}{\dfrac{n-1-2Ln(1-\alpha/2)}{\sum_{i=1}^{n}(x_i-\bar{x})^2} - \dfrac{1}{f_1+f_2} Ln\left(\dfrac{\sum_{i=1}^{n}(x_i-\bar{x})^2 + f_2}{\sum_{i=1}^{n}(x_i-\bar{x})^2 - f_1}\right)}.$$

5.3 Main Results

Using the above Approximate Bayesian confidence bounds for a Normal population variance along with

$$\sigma^2 = E(X^2) - \mu^2$$

, we can easily derive the following Approximate Bayesian confidence bounds for a strictly positive mean of a normal population:

$$L_{\mu(HT)} = \sqrt{s^2 + \bar{x}^2 - \cfrac{1}{\cfrac{n-1-2Ln(1-\alpha/2)}{\sum_{i=1}^{n}(x_i - \bar{x})^2} - \cfrac{1}{f_1+f_2} Ln\left(\cfrac{\sum_{i=1}^{n}(x_i - \bar{x})^2 + f_2}{\sum_{i=1}^{n}(x_i - \bar{x})^2 - f_1}\right)}}$$

$$U_{\mu(HT)} = \sqrt{s^2 + \bar{x}^2 - \cfrac{1}{\cfrac{n-1-2Ln(\alpha/2)}{\sum_{i=1}^{n}(x_i - \bar{x})^2} - \cfrac{1}{f_1+f_2} Ln\left(\cfrac{\sum_{i=1}^{n}(x_i - \bar{x})^2 + f_2}{\sum_{i=1}^{n}(x_i - \bar{x})^2 - f_1}\right)}}$$

Hence, for a normal random variable X with a mean that is smaller or equal to zero, we can infer the following Approximate Bayesian confidence bounds for the corresponding mean:

$$L_{\mu(HT)} = \left(s^2 + \bar{y}^2 - \cfrac{1}{\cfrac{n-1-2Ln(1-\alpha/2)}{\sum\limits_{i=1}^{n}(y-\bar{y})^2} - \cfrac{1}{f_1+f_2} Ln\left(\cfrac{\sum\limits_{i=1}^{n}(y_i-\bar{y})^2 + f_2}{\sum\limits_{i=1}^{n}(y_i-\bar{y})^2 - f_1}\right)} \right)^{0.5} - a$$

$$U_{\mu(HT)} = \left(s^2 + \bar{y}^2 - \cfrac{1}{\cfrac{n-1-2Ln(\alpha/2)}{\sum\limits_{i=1}^{n}(y-\bar{y})^2} - \cfrac{1}{f_1+f_2} Ln\left(\cfrac{\sum\limits_{i=1}^{n}(y_i-\bar{y})^2 + f_2}{\sum\limits_{i=1}^{n}(y_i-\bar{y})^2 - f_1}\right)} \right)^{0.5} - a$$

where y=x+a and "a" is a constant number such that x+a>0

5.4 Numerical results

Approximate Bayesian confidence intervals will be constructed with the use of samples that have been obtained from normally distributed populations (Examples 1, 2, 3, .4, 7) and approximately normal populations (Examples 5, 6). SAS software is used to obtain the normal population mean μ and standard deviation σ corresponding to each of the examples of Normal and approximately Normal data sets that are given below. The lengths of the classical and Approximate Bayesian confidence intervals are respectively denoted by l_C and l_{HT}.

Example 1

Data obtained from Prem. S. Mann, Introductory Statistics, Third edition, page 504, 1998

24, 28, 22, 25, 24, 22, 29, 26, 25, 28, 19, 29.

Normal population distribution obtained with SAS:

$$N(\mu = 25.083, \sigma = 3.1176)$$

The corresponding sample mean and sample variance are

$$\bar{x} = 25.08333$$

$$s^2 = 9.719696$$

Table 1: Classical and Approximate Bayesian confidence intervals for the population mean corresponding to the first example of data set.

C. L. %	Approximate Bayesian bounds (HT)	Classical bounds	WC WHT
80	25.0730-25.1158	23.85665-26.31001	57.32

90	25.0683-25.1311	23.46696-26.69971	51.48
95	25.0661-25.1437	23.10246-27.06420	51.05
99	25.0643-25.1660	22.28798-27.87869	54.97

Example 2

Data obtained from Prem. S. Mann, Introductory Statistics, Third edition, page 504, 1998

13, 11, 9, 12, 8, 10, 5, 10, 9, 12, 13.

Normal population distribution obtained with SAS:

$$N(\mu=10.182, \sigma=2.4008)$$

The corresponding sample mean and sample variance are

$$\bar{x} = 10.181812$$

$$s^2 = 5.763636$$

Table 2 Classical and Approximate Bayesian confidence intervals for the population mean corresponding to the second example of data set.

C.L. %	Approximate Bayesian bounds (HT)	Classical bounds	WC WHT
80	10.1652-10.2330	9.18869-11.17495	29.30
90	10.1575-10.2565	8.87019-11.49344	26.50
95	10.1538-10.2756	8.56907-11.79457	26.48

| 99 | 10.1506-10.3194 | 7.88792-12.47572 | 27.18 |

Example 3

Data obtained from Prem. S. Mann, Introductory Statistics, Third edition, page 504, 1998.

16, 14, 11, 19, 14, 17, 13, 16, 17, 18, 19, 12.

Normal population distribution obtained with SAS:

$N(\mu=15.5, \sigma=2.6799)$

The corresponding sample mean and sample variance are

$s^2 = 7.181818$

$\bar{x} = 15.5$

Table 3: Classical and Approximate Bayesian confidence intervals for the population mean corresponding to the third example of data set.

C.L. %	Approximate Bayesian bounds (HT)	Classical bounds	WC WHT
80	15.4877-15.5388	14.44556-16.55440	41.27
90	15.4820-15.5570	14.11058-16.88942	37.85
95	15.4794-15.5721	13.79727-17.20273	36.74
99	15.4773-15.5986	13.09714-17.90286	39.62

Example 4

Data obtained from Prem. S. Mann, Introductory Statistics, Third edition, page 504, 1998.

27, 31, 25, 33, 21, 35, 30, 26, 25, 31. 33. 30, 28.

Normal population distribution obtained with SAS:

$$N(\mu = 28.846, \sigma = 3.9549)$$

The corresponding sample mean and sample variance are

$$\bar{x} = 28.846153$$

$$s^2 = 15.641025$$

Table 4: Classical and Approximate Bayesian confidence intervals for the population mean corresponding to the fourth example of data set.

C. L. %	Approximate Bayesian bounds (HT)	Classical bounds	WC WHT
80	28.8330-28.8884	27.35878-30.33353	53.70
90	28.8270-28.9087	26.89151-30.80080	47.85
95	28.8242-28.9256	26.45604-31.2362	47.14
99	28.8220-28.9560	25.49517-32.19714	50.02

Example 5

Data obtained from James T. McClave/Terry Sincich, A first course in Statistics, page 301, Sixth edition, 1997

52, 33, 42, 44, 41, 50, 44, 51, 45, 38, 37, 40, 44, 50, 43.

ormal population distribution obtained with SAS:

$$N(\mu = 43.6, \sigma = 5.4746)$$

The corresponding sample mean and sample variance are

$$s^2 = 29.971428$$

$$\bar{x} = 43.6$$

Table 5: Classical and Approximate Bayesian confidence intervals for the population mean corresponding to the fifth example of data set.

C.L. %	Approximate Bayesian bounds (HT)	Classical bounds	WC WHT
80	43.5858-43.6169	41.69879-45.50121	122.3
90	43.5794-43.6703	41.11076-46,08924	54.77
95	43.5764-43.6902	40.56796-46.63204	53.29
99	43.5741-43.7268	39.39189-47.80811	55.12

Example 6

Data obtained from James T. McClave/Terry Sincich, A first course in Statistics, page 301, Sixth edition, 1997

52, 43, 47, 56, 62, 53, 61, 50, 56, 52, 53, 60, 50, 48, 60, 55.

Normal population distribution obtained with SAS:

$$N(\mu = 53.625, \sigma = 5.4145)$$

$s^2 = 29.316666$
The corresponding sample mean and sample variance are

$$\bar{x} = 53.625$$

Table 6: Classical and Approximate Bayesian confidence intervals for the population mean corresponding to the sixth example of data set.

C.L. %	Approximate Bayesian bounds (HT)	Classical bounds	WC WHT
80	53.6145-53.6602	51.80979-55.44021	79.44
90	53.6098-53.6779	51.25210-55.99790	69.69
95	53.6076-53.6932	50.74043-56.50957	67.40
99	53.6058-53.7216	49.63588-57.61412	68.90

Example 7

The following observations have been obtained from the collection of SAS data sets.

50, 65, 100, 45, 111, 32, 45, 28, 60, 66, 114, 134, 150, 120, 77, 108, 112, 113, 80, 77, 69, 91, 116, 122, 37, 51, 53, 131, 49, 69, 66, 46, 131, 103, 84, 78.

$N(\mu = 82.861, \sigma = 33.226)$

Normal population distribution obtained with SAS:

$s^2 = 1103.951587$

The corresponding sample mean and sample variance are

$\bar{x} = 82.8611$

Table 7: Classical and Approximate Bayesian confidence intervals for the population mean corresponding to the seventh example of data set.

C.L. %	Approiximate .Bayesian bounds (HT)	Classical bounds	WC WHT
80	82.7539-83.2572	75.6261-90.0959	28.75
90	82.7072-83.4808	73.5052-92.2168	24.19
95	82.6856-83.6884	71.6196-94.1024	22.42
99	82.6690-83.7173	67.7793-97.9427	28.77

The above tables show that the Approximate Bayesian confidence intervals contain the population mean and are strictly included in their classical counterparts; also, the widths of the classical confidence intervals are more than twenty-two times greater than the ones corresponding to the new Approximate Bayesian approach.

5.5 Summary and Conclusion

i) The classical method and the Approximate Bayesian approach perform well.

ii) The classical method used to constructing confidence intervals for the mean of a Normal population does not always yield the best coverage accuracy. In fact, the obtained Approximate Bayesian confidence intervals perform better than their classical counterparts.

iii) Contrary to the classical method that uses the standard Normal and the student-t statistics, the new Approximate Bayesian approach and confidence bounds rely only on the observations that are under study.

Chapter 6 Approximate Bayesian Confidence intervals for the Mean of a Gaussian Distribution Versus Bayesian models

6.1 Introduction

Chapter 6 is concerned with the sensitivity of Bayesian analysis with respect to the choice of the prior distribution.

In this Chapter, we will compare a published Bayesian model with the new Approximate Bayesian confidence intervals for a normal population mean that were obtained in Chapter 4 and Chapter 5.

6.2 Preliminaries

Employing the Square Error loss function along with a normal prior, the Bayesian model (Dr; M. Fogel -1991) has the following bounds for the mean of the Normal probability density function:

$$L_B = \frac{\mu_1 \sigma^2/n + \bar{x}\tau^2}{\tau^2 + \sigma^2/n} - Z_{\alpha/2} \frac{\tau\sigma/\sqrt{n}}{\sqrt{\tau^2 + \sigma^2/n}}$$

$$U_B = \frac{\mu_1 \sigma^2/n + \bar{x}\tau^2}{\tau^2 + \sigma^2/n} + Z_{\alpha/2} \frac{\tau\sigma/\sqrt{n}}{\sqrt{\tau^2 + \sigma^2/n}}$$

. where the mean and variance of the selected normal prior are respectively denoted by

μ :and $\tau^2{}_1$

6.3 Main results

In chapter 3, with the use of the Square Error loss function, we derived the following Approximate Bayesian Confidence bounds for the Mean of a Normal probability distribution:

$$L_{\mu(SE)} = \left(\frac{\sum_{i=1}^{n}(x_i - \bar{x})^2}{n-1} + \bar{x}^2 - \frac{\sum_{i=1}^{n}(x_i - \bar{x})^2}{n-2-2\ln(1-\alpha/2)} \right)^{0.5}$$

$$U_{\mu(SE)} = \left(\frac{\sum_{i=1}^{n}(x_i - \bar{x})^2}{n-1} + \bar{x}^2 - \frac{\sum_{i=1}^{n}(x_i - \bar{x})^2}{n-2-2\ln(\alpha/2)} \right)^{0.5}$$

When a normal random variable X has a mean that is smaller or equal to zero, we have the following Approximate Bayesian confidence bounds for the corresponding population mean:

$$U_{\mu(SE)} = \left(\frac{\sum_{i=1}^{n}(y_i - \bar{y})^2}{n-1} + \bar{y}^2 - \frac{\sum_{i=1}^{n}(y_i - \bar{y})^2}{n-2-2\ln(\alpha/2)} \right)^{0.5} - a$$

$$L_{\mu(SE)} = \left(\frac{\sum_{i=1}^{n}(y_i-\bar{y})^2}{n-1} + \bar{y}^2 - \frac{\sum_{i=1}^{n}(y_i-\bar{y})^2}{n-2-2\ln(1-\alpha/2)} \right)^{0.5} - a$$

where y=x+a and "a" is a constant such that x+a>0

With the Higgins-Tsokos Loss function, for a strictly positive mean of a normal population we have

$$U_{\mu(HT)} = \left(s^2 + \bar{x}^2 - \frac{1}{\frac{n-1-2Ln(\alpha/2)}{\sum_{i=1}^{n}(x_i-\bar{x})^2} - \frac{1}{f_1+f_2} Ln\left(\frac{\sum_{i=1}^{n}(x_i-\bar{x})^2 + f_2}{\sum_{i=1}^{n}(x_i-\bar{x})^2 - f_1}\right)} \right)^{0.5}$$

$$L_{\mu(HT)} = \left(s^2 + \bar{x}^2 - \frac{1}{\frac{n-1-2Ln(1-\alpha/2)}{\sum_{i=1}^{n}(x_i-\bar{x})^2} - \frac{1}{f_1+f_2} Ln\left(\frac{\sum_{i=1}^{n}(x_i-\bar{x})^2 + f_2}{\sum_{i=1}^{n}(x_i-\bar{x})^2 - f_1}\right)} \right)^{0.5}$$

For the Higgins-Tsokos loss function, we have the following confidence bounds for a Normal population mean, of a Normal random variable X, that is smaller or equal to zero::

$$L_{\mu(HT)} = \left(s^2 + \bar{y}^2 - \cfrac{1}{\cfrac{n-1-2Ln(1-\alpha/2)}{\sum_{i=1}^{n}(y-\bar{y})^2} - \cfrac{1}{f_1+f_2} Ln\left[\cfrac{\sum_{i=1}^{n}(y_i-\bar{y})^2 + f_2}{\sum_{i=1}^{n}(y_i-\bar{y})^2 - f_1}\right]} \right)^{0.5} - a$$

$$U_{\mu(HT)} = \left(s^2 + \bar{y}^2 - \cfrac{1}{\cfrac{n-1-2Ln(\alpha/2)}{\sum_{i=1}^{n}(y-\bar{y})^2} - \cfrac{1}{f_1+f_2} Ln\left[\cfrac{\sum_{i=1}^{n}(y_i-\bar{y})^2 + f_2}{\sum_{i=1}^{n}(y_i-\bar{y})^2 - f_1}\right]} \right)^{0.5} - a$$

Where y=x+a and "a" is a constant such that x+a>0

6.4 Numerical results

To obtain numerical results, we will use samples that have been drawn from normally distributed populations (Examples 1, 2, 3, .4, 7) and approximately normal populations (Examples 5, 6)

SAS software is used to obtain the normal population mean μ and standard deviation σ corresponding to each of the data sets.

For the Higgins-Tsokos lass function, we will consider f1=1, f2=1.

Example 1

Data obtained from Prem. S. Mann, Introductory Statistics, Third edition, page 504, 1998

24, 28, 22, 25, 24, 22, 29, 26, 25, 28, 19, 29.

Normal population distribution obtained with SAS:

$$N(\mu = 25.083, \sigma = 3.1176)$$

The corresponding sample mean and sample variance are

$$\bar{x} = 25.08333$$

$$s^2 = 9.719696$$

Table 1: Bayesian and Approximate Bayesian confidence intervals for the population mean corresponding to the first example of data set.

C. L. %	Approximate Bayesian bounds (SE)	Approx. Bayesian bounds (HT)
80	25.0683-25.1311	25.0730-25.1158
90	25.0661-25.1437	25.0683-25.1311
95	25.0650-25.1543	25.0661-25.1437
99	25.0641-25.1734	25.0643-25.1660

C. L. %	Bayesian C. I. I Bayesian bounds $\mu_1 = 2, \tau = 1$	Bayesian C. I. II Bayesian bounds $\mu_1 = 25, \tau = 10$
80	13.8971-15.6097	23.9353-26.2300
90	13.6496-15.8572	23.6037-26.5617
95	13.4422-16.0646	23.3258-26.8395
99	13.0275-16.4793	22.7701-27.3953

Example 2

Data obtained from Prem. S. Mann, Introductory Statistics, Third edition, page 504, 1998

13, 11, 9, 12, 8, 10, 5, 10, 9, 12, 13.

Normal population distribution obtained with SAS:

$$N(\mu = 10.182, \sigma = 2.4008)$$

The corresponding sample mean and sample variance are

$$\bar{x} = 10.181812$$

$$s^2 = 5.763636$$

Table 2 **Bayesian and Approximate Bayesian confidence intervals for the population mean corresponding to the second example of data set.**

C. L. %	Approximate Bayesian bounds (SE)	Approximate Bayesian bounds (HT)
80	10.1575-10.2565	10.1652-10.2330
90	10.1538-10.2756	10.1575-10.2565
95	10.1520-10.2914	10.1538-10.2756
99	10.1506-10.3194	10.1506-10.3194

C. L. %	Bayesian C. I. I	Bayesian C. I. II
	Bayesian bounds : $\mu_1 = 2, \tau = 1$	Bayesian bounds $\mu_1 = 25, \tau = 10$
80	6.6182-8.1193	9.3349-11.1832

90	6.4013-8.3363	9.0678-11.4503
95	6.2195-8.5180	8.8440-11.6741
99	5.8560-8.8816	8.3964-12.1217

Example 3

Data obtained from Prem. S. Mann, Introductory Statistics, Third edition, page 504, 1998.

16, 14, 11, 19, 14, 17, 13, 16, 17, 18, 19, 12.

Normal population distribution obtained with SAS:

$$N(\mu = 15.5, \sigma = 2.6799)$$

The corresponding sample mean and sample variance are

$$\bar{x} = 15.5$$

$$s^2 = 7.181818$$

Table 3: Bayesian and Approximate Bayesian confidence intervals for the population mean corresponding to the third example of data set.

C.L. %	Approximate Bayesian bounds (SE)	Approximate Bayesian bounds (HT)
80	15.4820-15.5570	15.4877-15.5388
90	15.4794-15.5721	15.4820-15.5570
95	15.4781-15.5847	15.4794-15.5721
99	15.4770-15.6075	15.4773-15.5986

C.L. %	Bayesian C. I. I Bayesian bounds: $\mu_1=2, \tau=1$	Bayesian C. I. II Bayesian bounds $\mu_1=25, \tau=10$
80	9.6623-11.2287	14.5692-16.5438
90	9.4359-11.4551	14.2839-16.8292
95	9.2462-11.6448	14.0447-17.0683
99	8,8668-12.0242	13.5665-17.5465

Example 4

Data obtained from Prem. S. Mann, Introductory Statistics, Third edition, page 504, 1998.

27, 31, 25, 33, 21, 35, 30, 26, 25,31.33.30, 28.

Normal population distribution obtained with SAS:

$N(\mu = 28.846, \sigma = 3.9549)$

The corresponding sample mean and sample variance are

$\bar{x} = 28.846153$

$s^2 = 15.641025$

Table 4: **Bayesian and Approximate Bayesian confidence intervals for the population mean corresponding to the fourth example of data set.**

C. L. %	Approximate Bayesian bounds (SE)	Approximate Bayesian bounds (HT)
80	28.8270-28.9087	28.8330-28.8884
90	28.8242-28.9256	28.8270-28.9087
95	28.8228-28.9400	28.8242-28.9256
99	28.8217-28.9663	28.8220-28.9560

C. L. %	Bayesian C. I. I Bayesian bounds: $\mu_1 = 2, \tau = 1$	Bayesian C. I. II Bayesian bounds $\mu_1 = 25, \tau = 10$
80	13.2394-15.1312	27.4048-30.1961
90	12.9659-15.4047	27.0014-30.5995

95	12.7369-15.6337	26.6634-30.9375
99	12.2787-16.0919	25.9873-31.6135

Example 5

Data obtained from James T. McClave/Terry Sincich, A first course in Statistics, page 301, Sixth edition, 1997

52, 33, 42, 44, 41, 50, 44, 51, 45, 38, 37, 40, 44, 50, 43.

$N(\mu = 43.6, \sigma = 5.4746)$

Normal population distribution obtained with SAS:

$\bar{x} = 43.6$

The corresponding sample mean and sample variance are

$s^2 = 29.971428$

Table 5: Bayesian and Approximate Bayesian confidence intervals for the population mean corresponding to the fifth example of data set.

C.L. %	Approximate Bayesian bounds (SE)	Approximate Bayesian bounds (HT)
80	43.5794-43.6703	43.5858-43.6169
90	43.5764-43.6902	43.5794-43.6703

95	43.5749-43.7074	43.5764-43.6902
99	43.5738-43.7395	43.5741-43.7268

C. L. %	Bayesian C. I. I Bayesian bounds: $\mu_1=2, \tau=1$	Bayesian C. I. II Bayesian bounds $\mu_1=25, \tau=10$
80	14.8305-16.9204	41.4441-45.0272
90	14.5285-17.2225	40.9263-45.5450
95	14.2754-17.4756	40.4924-45.9789
99	13.7692-17.9817	39.6246-46.8467

Example 6

Data obtained from James T. McClave/Terry Sincich, A first course in Statistics, page 301, Sixth edition, 1997

52, 43, 47, 56, 62, 53, 61, 50, 56, 52, 53, 60, 50, 48, 60, 55.

Normal population distribution obtained with SAS:

$$N(\mu=53.625, \sigma=5.4145)$$

The corresponding sample mean and sample variance are

$\bar{x} = 53.625$

$s^2 = 29.316666$

Table 6: Bayesian and Approximate Bayesian confidence intervals for the population mean corresponding to the sixth example of data set.

C. L. %	Approximate Bayesian bounds (SE)	Approximate Bayesian bounds (HT)
80	53.6098-53.6779	53.6145-53.6602
90	53.6076-53.6932	53.6098-53.6779
95	53.6065-53.7064	53.6076-53.6932
99	53.6056-53.7315	53.6058-53.7216

C. L. %	Bayesian C. I. I — Bayesian bounds: $\mu_1 = 2, \tau = 1$	Bayesian C. I. II — Bayesian bounds $\mu_1 = 25, \tau = 10$
80	19.1978-21.2568	51.3930-54.8269
90	18.9002-21.5544	50.8967-55.3232
95	18.6508-21.8038	50.4808-55.7391
99	18.1521-22.3024	49.6492-56.5707

Example 7

The following observations have been obtained from the collection of SAS data sets.

50, 65, 100, 45, 111, 32, 45, 28, 60, 66, 114, 134, 150, 120, 77, 108, 112, 113, 80, 77, 69, 91, 116, 122, 37, 51, 53, 131, 49, 69, 66, 46, 131, 103, 84, 78.

Normal population distribution obtained with SAS:

$$N(\mu = 82.861, \sigma = 33.226)$$

The corresponding sample mean and sample variance are

$$\bar{x} = 82.8611$$

$$s^2 = 1103.951587$$

Table 7: Bayesian and Approximate Bayesian confidence intervals for the population mean corresponding to the seventh example of data set.

C.L. %	Approximate Bayesian bounds (SE)	Approximate Bayesian bounds (HT)
80	82.7072-83.4808	82.7539-83.2572
90	82.6856-83.6884	82.7072-83.4808
95	82.6751-83.8815	82.6856-83.6884
99	82.6669-84.2823	82.6690-83.1173

C. L. %	Bayesian C. I. I Bayesian bounds : $\mu_1=2, \tau=1$	Bayesian C. I. II Bayesian bounds $\mu_1=25, \tau=10$
80	3.2940-5.8132	63.0810-75.4828
90	2.9299-6.17740	61.2886-77.2752
95	2.6248-6.4824	59.7868-78.7770
99	2.0147-7.0926	56.7833-81.7806

Contrary to the Approximate Bayesian models, the Bayesian confidence intervals do not always contain the population mean

6.5 Summary and Conclusions

In this Chapter, Approximate Bayesian confidence bounds for the mean of a normal population under two different loss functions have been compared to a published Bayesian model. The loss functions that are employed are the Square Error and the Higgins-Tsokos loss functions.

Based on the above numerical results we can conclude the following:

Bayesian models that are used to constructing confidence intervals for the mean of a Normal population do not always yield the best coverage accuracy. In fact, in the above numerical results, each of the obtained Approximate Bayesian confidence intervals contains the population mean, and performs better than its Bayesian counterparts

Bayesian models are in general sensitive to the choice of the hyper-parameters. Some values arbitrarily assigned to the hyper-parameters may lead to a very poor estimation of the parameter(s) under study,

Contrary to the above Bayesian model that uses the Z-table, the Approximate Bayesian approach and models rely only on the observations that are under study.

With the Approximate Bayesian Approach, confidence intervals for a normal population mean are easily obtained for any level of significance.

The Approximate Bayesian approach under the "popular" Square Error loss function does not always yield the best Approximate Bayesian results. In fact, in all seven examples, the Higgins-Tsokos loss function performs better.

Chapter 7 Approximate Bayesian Confidence Intervals For The Coefficient of Variation of a Gaussian Distribution corresponding to The Square Error loss function

7.1 Introduction

The aim of Chapter 7 is to derive closed-form confidence bounds for the coefficient of variation of a Gaussian distribution, with the use of the Square Error loss function.

We shall consider the Normal underlying model characterized by

$$f(x) = \frac{1}{\sqrt{2\pi}\sigma} e^{-\frac{1}{2}\left(\frac{x-\mu}{\sigma}\right)^2} ; -\infty \prec x \prec \infty, -\infty \prec \mu \prec \infty, \sigma \succ 0.$$

Considering the Square Error loss function, we will construct approximate Bayesian confidence intervals for the coefficients of variation of Normal populations.

The Approximate Bayesian results will then be compared to their classical counterparts corresponding to a published model (Miller E. G. -1991):

For a given level of significance, Miller's approach considers the following classical model, to construct confidence intervals for the coefficient of variation of a Normal population:

$$L_M = \frac{S}{\overline{X}} - Z_{\alpha/2} \left[m^{-1} \left(\frac{S}{\overline{X}}\right)^2 \left(0.5 + \left(\frac{S}{\overline{X}}\right)^2\right) \right]^{\frac{1}{2}}$$

$$U_M = \frac{S}{\overline{X}} + Z_{\alpha/2} \left[m^{-1} \left(\frac{S}{\overline{X}}\right)^2 \left(0.5 + \left(\frac{S}{\overline{X}}\right)^2\right) \right]^{\frac{1}{2}}$$

Where m=n-1, n being the sample size.

7.2 Preliminaries

To obtain the Approximate Bayesian confidence bounds for the coefficient of variation of a Normal population, we will use the following models that were obtained in Chapter 2 and Chapter 4:

Approximate Bayesian confidence bounds for the variance of a Gaussian distribution

$$L_{b(SE)} = \frac{\sum_{i=1}^{n}(x_i - \bar{x})^2}{n - 2 - 2\ln(\alpha/2)}$$

$$U_{b(SE)} = \frac{\sum_{i=1}^{n}(x_i - \bar{x})^2}{n - 2 - 2\ln(1 - \alpha/2)}$$

Approximate Bayesian confidence bounds for the mean of a Gaussian distribution

$$U_{\mu(SE)} = \left(\frac{\sum_{i=1}^{n}(x_i - \bar{x})^2}{n-1} + \bar{x}^2 - \frac{\sum_{i=1}^{n}(x_i - \bar{x})^2}{n-2-2\ln(\alpha/2)} \right)^{0.5}$$

$$L_{\mu(SE)} = \left(\frac{\sum_{i=1}^{n}(x_i - \bar{x})^2}{n-1} + \bar{x}^2 - \frac{\sum_{i=1}^{n}(x_i - \bar{x})^2}{n-2-2\ln(1-\alpha/2)} \right)^{0.5}$$

$$U_{\mu(SE)} = \left(\frac{\sum_{i=1}^{n}(y_i - \bar{y})^2}{n-1} + \bar{y}^2 - \frac{\sum_{i=1}^{n}(y_i - \bar{y})^2}{n-2-2\ln(\alpha/2)} \right)^{0.5} - a$$

For a normal random variable X with a mean that is smaller or equal to zero, we can infer the following Approximate Bayesian confidence bounds:

$$L_{\mu(SE)} = \left(\frac{\sum_{i=1}^{n}(y_i - \bar{y})^2}{n-1} + \bar{y}^2 - \frac{\sum_{i=1}^{n}(y_i - \bar{y})^2}{n-2-2\ln(1-\alpha/2)} \right)^{0.5} - a$$

where y=x+a and "a" is a constant such that x+a>0

7.3 Main Results

Using the Approximate Bayesian models for the mean and the variance of a Normal distribution, we can infer the following Approximate Bayesian confidence bounds for the coefficient of variation:

$$L_{Cv}(SE) = \frac{\sqrt{L_{\sigma^2(SE)}}}{U_{\mu(SE)}}$$

$$U_{Cv}(SE) = \frac{\sqrt{U_{\sigma^2(SE)}}}{L_{\mu(SE)}}$$

Hence, for a positive mean, we have the following confidence bounds for the coefficient of variation:

$$U_{Cv}(SE) = \left(\frac{n-1}{\left(1 + \left(\frac{\bar{x}}{s}\right)^2\right)(n-2-2Ln(1-\alpha/2)) - n + 1} \right)^{0.5}$$

$$L_{Cv}(SE) = \left(\frac{n-1}{\left(1+\left(\frac{\bar{x}}{s}\right)^2\right)(n-2-2Ln(\alpha/2))-n+1} \right)^{0.5}$$

7.4 Numerical Results

To assess the performance of the above Approximate Bayesian model, Approximate Bayesian confidence intervals will be constructed with the use of samples that have been drawn from normally distributed populations (Examples 1, 2, 3, .4, 7) and approximately normal populations (Examples 5, 6)

SAS software is used to obtain the Normal population mean μ and standard deviation σ corresponding to each of the data set.

WM and WSE correspond respectively to the widths of the classical and Approximate Bayesian confidence intervals.

Example 1

Data obtained from Prem. S. Mann, Introductory Statistics, Third edition, page 504, 1998

24, 28, 22, 25, 24, 22, 29, 26, 25, 28, 19, 29.

Normal population distribution obtained with SAS:

$$N(\mu = 25.083, \sigma = 3.1176)$$

The corresponding coefficient of variation is

$$\bar{C}_v = \frac{\sigma}{\mu} = 0.12429$$

Table 1: Miller's confidence bounds and Approximate Bayesian confidence intervals for the population coefficient of variation corresponding to the first data set.

C. L. %	Approx. Bayesian bounds (SE)	Miller's C. I.	WM / WSE
80	0.1077-0.1291	0.0899-0.1587	3.215
90	0.1028-0.1298	0.0799-0.1687	3.289
95	0.0986-0.1301	0.0716-0.1770	3.346
99	0.0905-0.1303	0.0549-0.1937	3.487

Example 2

Data obtained from Prem. S. Mann, Introductory Statistics, Third edition, page 504, 1998

13, 11, 9, 12, 8, 10, 5, 10, 9, 12, 13.

Normal population distribution obtained with SAS:

$$N(\mu=10.182, \sigma=2.4008)$$

The corresponding coefficient of variation is

$$\bar{C}_v = \frac{\sigma}{\mu} = 0.24579$$

Table 2 Miller's confidence bounds and Approximate Bayesian confidence intervals for the population coefficient of variation corresponding to the second example of data set.

C.L. %	Approx. Bayesian bounds (SE)	Miller's C. I.	WM / WSE
80	0.2007-0.2463	0.1647-0.3069	3.118
90	0.1908-0.2478	0.1441-0.3275	3.218
95	0.1823-0.2486	0.1269-0.3447	3.285
99	0.1662-0.2492	0.0924-0.3792	3.455

Example 3

Data obtained from Prem. S. Mann, Introductory Statistics, Third edition, page 504, 1998.

16, 14, 11, 19, 14, 17, 13, 16, 17, 18, 19, 12.

Normal population distribution obtained with SAS:

$$N(\mu=15.5, \sigma=2.6799)$$

The corresponding coefficient of variation is

$$\bar{C}_v = \frac{\sigma}{\mu} = 0.17290$$

Table 3: Miller's confidence bounds and Approximate Bayesian confidence intervals for the population coefficient of variation corresponding to the third example of data set.

C. L. %	Approx. Bayesian bounds (SE)	Miller's C. I.	WM WSE
80	0.1495-0.1797	0.1243-0.2215	3.219
90	0.1427-0.1807	0.1103-0.2354	3.292
95	0.1368-0.1811	0.0985-0.2473	3.359
99	0.1255-0.1815	0.0750-0.2708	3.496

Example 4

Data obtained from Prem. S. Mann, Introductory Statistics, Third edition, page 504, 1998.

27, 31, 25, 33, 21, 35, 30, 26, 25, 31. 33. 30, 28.

Normal population distribution obtained with SAS:

$$N(\mu=28.846, \sigma=3.9549)$$

$$\bar{C}_v = \frac{\sigma}{\mu} = 0.13710$$

The corresponding coefficient of variation is

Table 4: Miller's confidence bounds and Approximate Bayesian confidence intervals for the population coefficient of variation corresponding to the fourth example of data set.

C.L. %	Approx. Bayesian bounds (SE)	Miller's C. I.	WM WSE
80	0.1200-0.1419	0.1006-0.1736	3.333
90	0.1150-0.1426	0.0901-0.18414	3.406
95	0.1104-0.1430	0.0812-0.1930	3.429
99	0.1018-0.1433	0.0636-0.2107	3.545

Example 5

Data obtained from James T. McClave/Terry Sincich, A first course in Statistics, page 301, Sixth edition, 1997

52, 33, 42, 44, 41, 50, 44, 51, 45, 38, 37, 40, 44, 50, 43.

Normal population distribution obtained with SAS:

$$N(\mu = 43.6, \sigma = 5.4746)$$

The corresponding coefficient of variation is

$$\bar{C}_v = \frac{\sigma}{\mu} = 0.13710$$

Table 5: Miller's confidence bounds and Approximate Bayesian confidence intervals for the population coefficient of variation corresponding to the fifth example of data set.

C. L. %	Approx.Bayesian bounds (SE)	Miller's C. I.	WM WSE
80	0.1118-0.1293	0.09472-0.1564	3.526
90	0.1076-0.1299	0.08580-0.1653	3.565
95	0.1039-0.1301	0.0783-0.1728	3.607
99	0.0964-0.1303	0.0634-0.1877	3.667

Example 6

Data obtained from James T. McClave/Terry Sincich, A first course in Statistics, page 301, Sixth edition, 1997

52, 43, 47, 56, 62, 53, 61, 50, 56, 52, 53, 60, 50, 48, 60, 55.

Normal population distribution obtained with SAS:

$$N(\mu = 53.625, \sigma = 5.4145)$$

The corresponding coefficient of variation is

$$\bar{C}_v = \frac{\sigma}{\mu} = 0.10097$$

Table 6: Miller's confidence bounds and Approximate Bayesian confidence intervals for the population coefficient of variation corresponding to the sixth example of data set.

C. L. %	Approx.Bayesian bounds (SE)	Miller's C. I.	WM WSE
80	0.0906-0.1038	0.0771-0.1248	3.614
90	0.0874-0.1042	0.0702-0.1317	3.661
95	0.0845-0.1044	0.0645-0.1375	3.668
99	0.0787-0.1045	0.0529-0.1490	3.725

Example 7

The following observations have been obtained from the collection of SAS data sets.

50, 65, 100, 45, 111, 32, 45, 28, 60, 66, 114, 134, 150, 120, 77, 108, 112, 113, 80, 77, 69, 91, 116, 122, 37, 51, 53, 131, 49, 69, 66, 46, 131, 103, 84, 78.

$$N(\mu = 82.861, \sigma = 33.226)$$

Normal population distribution obtained with SAS:

The corresponding coefficient of variation is

$$\bar{C}_v = \frac{\sigma}{\mu} = 0.40098$$

Table 7: Miller's confidence bounds and Approximate Bayesian confidence intervals for the population coefficient of variation corresponding to the seventh example of data set.

C.L. %	Approx. Bayesian bounds (SE)	Millre's C. I.	WM WSE
80	0.3790-0.4063	0.3305-0.4715	5.165
90	0.3714-0.4071	0.3101-0.4919	5.092
95	0.3643-0.4075	0.2930-0.5090	5.000
99	0.3492-0.4077	0.2588-0.5431	4.860

Each of the above Tables shows that the Approximate Bayesian confidence intervals contains the population coefficient of variation. The Approximate Bayesian confidence intervals are also strictly included in their classical counterparts.

7.5 Summary and Conclusions

In this Chapter, Approximate Bayesian confidence intervals for a coefficient of variation of a Normal population have been constructed with the Square Error loss function.

Based on the above numerical results we can conclude the following:

The new Approximate Bayesian confidence bounds rely only on the observations that are under study; they also perform well and have great coverage accuracy.

Miller's approach used to constructing confidence intervals for the coefficient of variation of a normal population does not always yield the best coverage accuracy. In fact, the obtained approximate Bayesian confidence intervals perform better than their counterpart obtained with Miller's approach

With the Approximate Bayesian approach, confidence intervals for a Normal population coefficient of variation are easily obtained for any level of significance and any sample size.

Chapter 8 Approximate Bayesian Confidence Intervals For The Coefficient of Variation of a Gaussian Distribution corresponding to The Higgins-Stokes loss function

8.1 Introduction

In Chapter 8, we will derive a closed-form Approximate Bayesian confidence bounds for the coefficient of variation of a Normal population, with the use of the Higgins-Tsokos loss function.

We shall consider the Normal underlying model characterized by

$$f(x) = \frac{1}{\sqrt{2\pi}\sigma} e^{-\frac{1}{2}\left(\frac{x-\mu}{\sigma}\right)^2} ; -\infty \prec x \prec \infty, -\infty \prec \mu \prec \infty, \sigma \succ 0.$$

Considering the Higgins-Tsokos loss function, we will construct Approximate Bayesian confidence bounds for the coefficients of variation of a Normal population.

To assess the performance of the Approximate Bayesian models, numerical results will be obtained with the use of Normal and Approximately Normal data along with SAS software. The Approximate Bayesian results will then be compared to their classical counterparts corresponding to the following a model (Miller E. G. 1991):

$$L_M = \frac{S}{\overline{X}} - Z_{\alpha/2}\left[m^{-1}\left(\frac{S}{\overline{X}}\right)^2\left(0.5 + \left(\frac{S}{\overline{X}}\right)^2\right)\right]^{\frac{1}{2}}$$

$$U_M = \frac{S}{\overline{X}} + Z_{\alpha/2}\left[m^{-1}\left(\frac{S}{\overline{X}}\right)^2\left(0.5 + \left(\frac{S}{\overline{X}}\right)^2\right)\right]^{\frac{1}{2}}$$

Where m=n-1, n being the sample size.

8.2 Preliminaries

We will use the following models that were obtained in Chapter 3 and Chapter 5:

Approximate Bayesian confidence bounds for the variance of a Gaussian distribution.

$$L_{b(HT)} = \dfrac{1}{\dfrac{n-1-2Ln(\alpha/2)}{\sum_{i=1}^{n}(x_i - \bar{x})^2} - \dfrac{1}{f_1 + f_2} Ln\left(\dfrac{\sum_{i=1}^{n}(x_i - \bar{x})^2 + f_2}{\sum_{i=1}^{n}(x_i - \bar{x})^2 - f_1}\right)}$$

$$U_{b(HT)} = \dfrac{1}{\dfrac{n-1-2Ln(1-\alpha/2)}{\sum_{i=1}^{n}(x_i - \bar{x})^2} - \dfrac{1}{f_1 + f_2} Ln\left(\dfrac{\sum_{i=1}^{n}(x_i - \bar{x})^2 + f_2}{\sum_{i=1}^{n}(x_i - \bar{x})^2 - f_1}\right)}$$

Approximate Bayesian confidence bounds for a positive mean of a Gaussian distribution

$$L_{\mu(HT)} = \left(s^2 + \bar{x}^2 - \cfrac{1}{\cfrac{n-1-2Ln(1-\alpha/2)}{\sum_{i=1}^{n}(x_i-\bar{x})^2} - \cfrac{1}{f_1+f_2} Ln\left(\cfrac{\sum_{i=1}^{n}(x_i-\bar{x})^2 + f_2}{\sum_{i=1}^{n}(x_i-\bar{x})^2 - f_1}\right)} \right)^{0.5}$$

$$U_{\mu(HT)} = \left(s^2 + \bar{x}^2 - \cfrac{1}{\cfrac{n-1-2Ln(\alpha/2)}{\sum_{i=1}^{n}(x_i-\bar{x})^2} - \cfrac{1}{f_1+f_2} Ln\left(\cfrac{\sum_{i=1}^{n}(x_i-\bar{x})^2 + f_2}{\sum_{i=1}^{n}(x_i-\bar{x})^2 - f_1}\right)} \right)^{0.5}$$

8.3 Main results

To derive the Approximate Bayesian confidence bounds for the coefficient of variation of a Normal population, we will use the e Approximate Bayesian confidence bounds for the variance and the mean of a Gaussian distribution.

Hence, we have

$$L_{Cv}(SE) = \frac{\sqrt{L_{\sigma^2(HT)}}}{U_{\mu(HT)}}$$

$$U_{Cv}(SE) = \frac{\sqrt{U_{\sigma^2(HT)}}}{L_{\mu(HT)}} =$$

Thus, the Approximate Bayesian confidence bounds for a positive coefficient of variation of a Normal population are

$$L_{Cvr}(HT) = \frac{1}{\left(\left(s^2 + \bar{x}^2\right)\left(\frac{n-1-2Ln(\alpha/2)}{\sum_{i=1}^{n}(x_i - \bar{x})^2} - \frac{1}{f_1 + f_2} Ln\left(\frac{\sum_{i=1}^{n}(x_i - \bar{x})^2 + f_2}{\sum_{i=1}^{n}(x_i - \bar{x})^2 - f_1}\right)\right) - 1\right)^{0.5}}$$

$$U_{Cvr}(HT) = \frac{1}{\left(\left(s^2 + \bar{x}^2\right)\left(\frac{n-1-2Ln(1-\alpha/2)}{\sum_{i=1}^{n}(x_i - \bar{x})^2} - \frac{1}{f_1 + f_2} Ln\left(\frac{\sum_{i=1}^{n}(x_i - \bar{x})^2 + f_2}{\sum_{i=1}^{n}(x_i - \bar{x})^2 - f_1}\right)\right) - 1\right)^{0.5}}$$

8.4 Numerical Results

To assess the performance of the above Approximate Bayesian confidence bounds, Approximate Bayesian confidence intervals will be constructed with the use of samples that have been drawn from normally distributed populations (Examples 1, 2, 3, .4, 7) and approximately normal populations (Examples 5, 6).

SAS software is employed to obtain the normal population mean μ and standard deviation σ corresponding to each of the data sets that are given below. We will consider f1=1, f2=1.WM and WHT correspond respectively to the widths of the classical and Approximate Bayesian confidence intervals.

Example 1

Data obtained from Prem. S. Mann, Introductory Statistics, Third edition, page 504, 1998

24, 28, 22, 25, 24, 22, 29, 26, 25, 28, 19, 29.

Normal population distribution obtained with SAS:

$$N(\mu = 25.083, \sigma = 3.1176)$$

The corresponding coefficient of variation is

$$\bar{C}_v = \frac{\sigma}{\mu} = 0.12429$$

Table 1: Miller's confidence bounds and Approximate Bayesian confidence intervals for the population coefficient of variation corresponding to the first example of data set.

C.L. %	Approx.Bayesian bounds (HT)	Miller's C. I.	WM WHT
80	0.1132-0.1276	0.0899-0.1587	4.778
90	0.1077-0.1291	0.0799-0.1687	4.150

| 95 | 0.1028-0.1298 | 0.0716-0.1770 | 3.904 |
| 99 | 0.0937-0.1303 | 0.0549-0.1937 | 3.792 |

Example 2

Data obtained from Prem. S. Mann, Introductory Statistics, Third edition, page 504, 1998

13, 11, 9, 12, 8, 10, 5, 10, 9, 12, 13.

$$N(\mu = 10.182, \sigma = 2.4008)$$

Normal population distribution obtained with SAS:

The corresponding coefficient of variation is

$$\bar{C}_v = \frac{\sigma}{\mu} = 0.24579$$

Table 2 Miller's confidence bounds and Approximate Bayesian confidence intervals for the population coefficient of variation corresponding to the second example of data set.

C.L. %	Approximate Bayesian bounds (HT)	Miller's C. I.	WM WHT
80	0.2122-0.2430	0.1647-0.3069	4.617

90	0.2007-0.2463	0.1441-0.3275	4.022
95	0.1908-0.2478	0.1269-0.3447	3.821
99	0.1724-0.2490	0.0924-0.3792	3.744

Example 3

Data obtained from Prem. S. Mann, Introductory Statistics, Third edition, page 504, 1998.

16, 14, 11, 19, 14, 17, 13, 16, 17, 18, 19, 12.

$N(\mu = 15.5, \sigma = 2.6799)$

Normal population distribution obtained with SAS:

The corresponding coefficient of variation is

$$\bar{C}_v = \frac{\sigma}{\mu} = 0.17290$$

Table 3: **Miller's confidence bounds and Approximate Bayesian confidence intervals for the population coefficient of variation corresponding to the third example of data set.**

C. L. %	Approximate Bayesian bounds (HT)	Miller's C. I.	WM WHT
80	0.1573-0.1776	0.1243-0.2215	4.788
90	0.1495-0.1797	0.1103-0.2354	4.142

| 95 | 0.14273-0.1807 | 0.0985-0.2473 | 3.916 |
| 99 | 0.1300-0.1842 | 0.0750-0.2708 | 3.613 |

Example 4

Data obtained from Prem. S. Mann, Introductory Statistics, Third edition, page 504, 1998.

27, 31, 25, 33, 21, 35, 30, 26, 25,31.33.30, 28.

Normal population distribution obtained with SAS:

$$N(\mu = 28.846, \sigma = 3.9549)$$

The corresponding coefficient of variation is

$$\bar{C}_v = \frac{\sigma}{\mu} = 0.13710$$

Table 4: Miller's confidence bounds and Approximate Bayesian confidence intervals for the population coefficient of variation corresponding to the fourth example of data set.

C.L. %	Approximate Bayesian bounds (HT)	Miller's C. I.	WM WHT
80	0.1258-0.1404	0.1006-0.1736	5.000
90	0.1200-0.1419	0.0901-	4.292

		0.18414	
95	0.1149-0.1426	0.0812-0.1930	4.036
99	0.1052-0.1432	0.0636-0.2107	3.871

Example 5

Data obtained from James T. McClave/Terry Sincich, A first course in Statistics, page 301, Sixth edition, 1997

52, 33, 42, 44, 41, 50, 44, 51, 45, 38, 37, 40, 44, 50, 43.

Normal population distribution obtained with SAS:

$$N(\mu = 43.6, \sigma = 5.4746)$$

$$\bar{C}_v = \frac{\sigma}{\mu} = 0.13710$$

The corresponding coefficient of variation is

Table 5: Miller's confidence bounds and Approximate Bayesian confidence intervals for the population coefficient of variation corresponding to the fifth example of data set.

C.L. %	Approximate Bayesian bounds (HT)	Miller's C. I.	WM / WHT
80	0.1166-0.1282	0.09472-0.1564	5.314
90	0.1118-0.1293	0.08580-0.1653	4.543

| 95 | 0.1076-0.1299 | 0.0783-0.1728 | 4.238 |
| 99 | 0.0994-0.1303 | 0.0634-0.1877 | 4.023 |

Example 6

Data obtained from James T. McClave/Terry Sincich, A first course in Statistics, page 301, Sixth edition, 1997

52, 43, 47, 56, 62, 53, 61, 50, 56, 52, 53, 60, 50, 48, 60, 55.

Normal population distribution obtained with SAS:

$$N(\mu = 53.625, \sigma = 5.4145)$$

The corresponding coefficient of variation is

$$\bar{C}_v = \frac{\sigma}{\mu} = 0.10097$$

Table 6: Miller's confidence bounds and Approximate Bayesian confidence intervals for the population coefficient of variation corresponding to the sixth example of data set.

C.L. %	Approximate .Bayesian bounds (HT)	Miller's C. I.	WM WHT
80	0.0942-0.1029	0.0771-0.1248	5.483
90	0.0906-0.1038	0.0702-0.1317	4.659

| 95 | 0.0874-0.1042 | 0.0645-0.1375 | 4.345 |
| 99 | 0.0810-0.1045 | 0.0529-0.1490 | 4.089 |

Example 7

The following observations have been obtained from the collection of SAS data sets.

50, 65, 100, 45, 111, 32, 45, 28, 60, 66, 114, 134, 150, 120, 77, 108, 112, 113, 80, 77, 69, 91, 116, 122, 37, 51, 53, 131, 49, 69, 66, 46, 131, 103, 84, 78.

Normal population distribution obtained with SAS:

$$N(\mu = 82.861, \sigma = 33.226)$$

The corresponding coefficient of variation is

$$\bar{C}_v = \frac{\sigma}{\mu} = 0.40098$$

Table 7: Miller's confidence bounds and Approximate Bayesian confidence intervals for the population coefficient of variation corresponding to the seventh example of data set.

C.L. %	Approximate Bayesian bounds (HT)	Miller's C. I.	WM WHT
80	0.3870-0.4047	0.3305-0.4715	7.966
90	0.3790-0.4063	0.3101-0.4919	6.660
95	0.3714-0.4071	0.2930-0.5090	6.050
99	0.3598-0.4077	0.2588-0.5431	5.935

Each of the above Tables shows that the Approximate Bayesian confidence intervals contain the population coefficient of variation and are strictly included in their classical counterpart.

8.5 Summary and Conclusions

In this Chapter, new Approximate Bayesian confidence bounds for a coefficient of variation of a Normal population have been constructed, with the Higgins-Tsokos loss function.

Based on the above numerical results we can conclude the following:

The classical and Approximate Bayesian confidence intervals have great coverage accuracy; the Approximate Bayesian intervals perform better.

The new Approximate Bayesian confidence bounds rely only on the observations that are under study.

With the Approximate Bayesian approach, confidence intervals for a Normal population coefficient of variation are easily obtained for any level of significance and any sample size.

Chapter 9 Approximate Bayesian Confidence Bounds For The Mean of an Exponential Distribution corresponding to the Square Error loss function

9.1 Introduction

The aim of this Chapter is to construct closed-form confidence bounds for the mean of an Exponential distribution.

We shall consider the Exponential underlying model characterized by

$$f(x) = \theta e^{-\theta x}; x \geq 0, \theta \succ 0$$

With $\theta = \dfrac{1}{b}$ where "b" is the scale parameter.

Considering the Square loss function, we will construct new Approximate Bayesian confidence intervals for the mean of an Exponential distribution.

To assess the performance of the Approximate Bayesian confidence bounds, numerical results will be obtained with the use of Exponential data.

The Approximate Bayesian results will then be compared to their classical counterpart corresponding to the Fisher Matrix bounds method.

Once the underlying model is found to have an Exponential distribution, Fisher Matrix bounds method uses the Z-table and considers the following confidence bounds for θ:

$$L_\theta = \frac{\hat{\theta}}{e^{K_\alpha \frac{\sqrt{Var(\hat{\theta})}}{\hat{\theta}}}}$$

$$U_\theta = \hat{\theta} e^{K_\alpha \frac{\sqrt{Var(\hat{\theta})}}{\hat{\theta}}}$$

where K_α is defined by

$$\alpha = \frac{1}{\sqrt{2\pi}} \int_{K_\alpha}^{\infty} e^{-\frac{t}{2}} dt = 1 - \Phi(K_\alpha)$$

$$Var(\hat{\theta}) = \left(\frac{\partial^2 \Lambda}{\partial \theta^2}\right)^{-1}$$

9.2 Preliminaries

The Square Error loss is defined as follows :

$$L_{SE}(\hat{\theta}, \theta) = \left(\hat{\theta} - \theta\right)^2$$

We shall assume that θ behaves as a random variable that is characterized by the following Pareto probability density function:

$$f_1(\theta) = \frac{a}{b}\left(\frac{b}{\theta}\right)^{a+1}; \theta \geq b \succ 0, a \succ 0.$$

The Pareto prior has been selected because of its mathematical tractability.

Using Exponential data, we will approximate the above Pareto prior in such a way that good approximate Bayesian estimates of θ are obtained.

9.3 Main results.

Let $x_1, x_2, \ldots, x_n$ denote Exponential data.

We have the following posterior distribution:

$$h(\theta \backslash x) \frac{\theta^{n-a-1} e^{-\theta \sum_{1}^{n} x_i}}{\int_{b}^{\infty} \theta^{n-a-1} e^{-\theta \sum_{1}^{n} x_i} d\theta}, \theta \succ b..$$

With the following approximate prior, good approximate Bayesian estimates of θ are obtained:

$$a = n+1, \quad b = \frac{n-1}{\sum_{1}^{n} x_i}$$

.

$$f_1(\theta)=\frac{a}{b}\left(\frac{b}{\theta}\right)^{a+1} ; \theta \geq b \succ 0, a \succ 0.$$

It's easily shown that the Approximate Bayesian estimator of the parameter θ, subject to the above prior is equal to

$$\frac{n}{\sum_{i=1}^{n} x_i}$$

Using the corresponding approximate posterior distribution, along with the equalities $P(\theta \succ L| x) = 1 - \alpha/2$ and $P(\theta \succ U| x) = \alpha/2$, we obtain the following lower and upper confidence bounds for θ:

$$L_{SE} = \frac{n-1-\ln(1-\alpha/2)}{\sum_{1}^{n} x_i} \qquad U_{SE} = \frac{n-1-\ln(\alpha/2)}{\sum_{1}^{n} x_i}$$

Thus, the $100(1-\alpha)\%$ Approximate Bayesian confidence bounds for the scale parameter b are

$$L_{b(SE)} = \frac{\sum_{i=1}^{n} x_i}{n-1-\ln(\alpha/2)} \qquad U_{b(SE)} = \frac{\sum_{i=1}^{n} x_i}{n-1-\ln(1-\alpha/2)}$$

The $100(1-\alpha)\%$ Approximate Bayesian confidence bounds for the mean of the Exponential distribution are

$$L_{b(SE)} = \frac{\sum_{i=1}^{n} x_i}{n-1-\ln(\alpha/2)} \qquad U_{b(SE)} = \frac{\sum_{i=1}^{n} x_i}{n-1-\ln(1-\alpha/2)}$$

9.4 Numerical Results

Approximate Bayesian and Fisher Matrix bounds confidence intervals will be constructed with the use of Exponential data

The lengths of the Fisher Matrix and Approximate Bayesian confidence intervals are respectively denoted by l_F and l_{SE}.

Example 1

Monte Carlo simulation has been used to generate the following 30 observations from the Exponential distribution with mean equal to 9

2.0270, 4.0103, 30.0421, 0.1189, 2.7558. 13.7441, 13.3840, 27.0930, 7.3750, 3.7323, 23.4171, 0.06310. 5.6839, 8.7473, 10.2778, 25.2331, 10.1903, 0.3761, 3.3068, 3.4954, 6.9136, 1.8234, 16.3160, 2.4359, 19.9108, 2.5285, 3.9314, 3.4645, 6.9229, 10.4509.

Table 2: Fisher Matrix bounds and Approximate Bayesian confidence intervals for the exponential population mean. when the population mean is equal to 9.

Confidence level	Fisher Matrix bounds	Approx.Bayesian bounds (SE)
80%	7.1184 – 11.3598	8.6182 – 9.2688

90%	6.6534 – 12.1537	8.4315 – 9.2861
95%	6.2873 – 12.8614	8.2527 – 9.2944
99%	5.6144 – 14.4028	7.8655 – 9.3009

Confidence level	$(l_F) \div (l_{SE})$
80%	*6.5192*
90%	*6.4361*
95%	*6.3109*
99%	*6.1226*

Example 2

Monte Carlo simulation has been used to generate the following 40 observations from the exponential distribution with mean equal to 20

4.5046, 8.9119, 66.7603, 0.2643, 6.1241, 30.5425, 29.7423, 60.2067, 16.3891, 8.2941, 52.0380, 0.1402, 12.6309, 19.4385, 22.8395, 52.3378, 3.4389, 19.3268, 8.2350, 3.4737, 56.0736, 22.6451, 0.8359, 7.3484, 7.7675, 15.3635, 4.05222, 36.2578, 5.6189, 8.7365, 7.6990, 15.3844, 23.2242, 11.8542, 63.6975, 14.8772, 32.9585, 2.2127, 5,4132, 44.2462

Table 3: Fisher Matrix bounds and Approximate Bayesian confidence intervals for the exponential population mean when the population mean is equal to 20.

Confidence level	Fisher Matrix bounds	*Approx. Bayesian bounds (SE)*

80%	16.5786 – 24.8507	19.6574 – 20.7619
90%	15.6366 – 26.3479	19.3330 – 20.7907
95%	14.8886 – 27.6715	19.0191 – 20.8045
99%	13.4983 – 30.5216	18.3281 – 20.8153

Confidence level	$(l_F) \div (l_{SE})$
80%	7.4894
90%	7.3480
95%	7.1596
99%	6.8443

Example 3

The following exponential data were obtained by Washington State Department of Ecology while conducting research on the amount of lead concentration in certain types of fish found in the Spokane River.

Lead (Pb) Concentrations in 1999 Spokane River Fish			
Source: WA State Dept. of Ecology report			
concentrations in parts per million (ppm)			
Filets	Trout	Whitefish	Sucker
	0.480	0.020	0.088
	0.071	0.020	0.210
	0.110	0.020	0.280

	0.320	0.020	0.030
	0.120	0.020	0.036
	0.220	0.065	0.047
	0.055	0.020	0.077
	0.320	0.037	0.069
	0.077	0.020	0.160
	0.081	0.036	0.088
	0.170		0.120
	0.130		0.054
	0.110		0.080
	0.081		0.059
	0.098		0.094
	0.180		0.059
	0.230		0.068
	0.082		0.020
	0.210		0.090
	0.200		0.046
	0.025		
	0.038		
Mean	0.155	0.028	0.089
std dev	0.110	0.015	0.063

Table 4: Fisher Matrix bounds and Approximate Bayesian confidence intervals for the mean lead concentration in trout.

Confidence level	Fisher Matrix bounds	Approx. Bayesian bounds (SE)
80%	0.11791 - 0.20351	0.15280 – 0.169507
90%	0.10896 – 0.22021	0.14820 – 0.16996
95%	0.10199 – 0.23526	0.14386 – 0.17018
99%	0.08936 – 0.26851	0.13471 – 0.17035

Confidence level	$(l_F) \div (l_{SE})$
80%	5.1236
90%	5.1125
95%	5.0634
99%	5.0266

Table 5.: Fisher Matrix bounds and Approximate Bayesian confidence intervals for the mean lead concentration in whitefish.

Confidence level	Fisher Matrix bounds	Approx. Bayesian bounds (SE)
80%	0.01854 – 0.04167	0.02698 - 0.03429
90%	0.01649 – 0.04684	0.02528 – 0.03452
95%	0.01495 – 0.05166	0.02378 – 0.03464

99%	0.01229 – 0.06285	0.02090 – 0.03472

Confidence level	$(l_F) \div (l_{SE})$
80%	3.1641
90%	3.2846
95%	3.3802
99%	3.6584

Table 6: Fisher Matrix bounds and Approximate Bayesian confidence intervals for the mean lead concentration in sucker.

Confidence level	Fisher Matrix bounds	Approx. Bayesian bounds (SE)
80%	0.06666 – 0.11816	0.08742 – 0.09803
90%	0.06136 – 0.12835	0.08454 – 0.09833
95%	0.05725 – 0.13756	0.08183 – 0.09847
99%	0.04984 – 0.15802	0.07618 – 0.09858

Confidence level	$(l_F) \div (l_{SE})$

80%	4.8539
90%	4.8578
95%	4.8263
99%	4.8294

Example 5

The following exponential data represent a random sample of cycles to failure in ten-thousands for twenty heater switches subject to an overload voltage (Abdulaziz Elfessi & David M. Raineke ,2001)

0.01, 0.034, 0.194, 0.567, 0.601, 0.712, 1.291, 1.367, 1.949, 2.37, 2.411, 2.875, 3.162, 3.28, 3.491, 3.686, 3.854, 4.211, 4.397, 6.473.

They conducted some studies on the above data and obtained the following the following maximum likelihood estimate and 95% confidence interval for the parameter θ: 0.4261 and (0.2603 , 0.6322).

Table 7: Fisher Matrix bounds and approximate Bayesian confidence intervals for θ

Confidence level	Fisher Matrix bounds	*Approx.Bayesian bounds (SE)*
80%	0.32005 – 0.56732	0.38575 – 0.43256
90%	0.29464 – 0.61626	0.38460 – 0.44733
95%	0.27491 – 0.66049	0.38404 – 0.46210
99%	0.23932 – 0.75871	0.38361 – 0.49639

Confidence level	$(l_F) \div (l_{SE})$

80%	5.2824
90%	5.1270
95%	4.9353
99%	4.6053

9.5 Summary and Conclusions

In this Chapter, Approximate Bayesian confidence bounds for the inverse of the scale parameter and the mean of an exponential distribution have been constructed. The loss function that is employed is the Square Error loss function.

Based on the above numerical results we can conclude the following:

The classical and Approximate Bayesian models perform well; all the confidence intervals that have been constructed have great coverage accuracy.

The Fisher Matrix bounds method used to constructing confidence intervals for the inverse of the scale parameter and the mean of an exponential population does not always yield the best coverage accuracy. The Approximate approach and models perform often better.

Contrary to Fisher Matrix bounds method that uses the Z-table, the Approximate Bayesian approach and confidence intervals rely only on the observations that are under study.

With the Approximate Bayesian Approach, Approximate Bayesian confidence intervals for an Exponential population mean are easily computed for any level of significance.

Chapter 10 Approximate Bayesian Confidence Intervals For The Mean of an Exponential Distribution corresponding to the Higgins-Tsokos loss function

10.1 Introduction

In Chapter 10, we will construct and compare confidence intervals for the mean of an Exponential distribution. Considering the Higgins-Tsokos loss function, Approximate Bayesian confidence bounds for the mean of an Exponential population are derived. Using Exponential data, the obtained Approximate Bayesian confidence intervals will then be compared to their counterparts corresponding to the Fisher Matrix bounds method.

The exponential probability is defined by

$$f(x) = \theta e^{-\theta x}; x \geq 0, \theta \succ 0$$

With $\theta = \dfrac{1}{b}$ where "b" is the scale parameter.

As mentioned in Chapter 9, once the underlying model is found to have an Exponential distribution, Fisher Matrix bounds method uses the following model to construct confidence intervals for θ

$$L_\theta = \dfrac{\hat{\theta}}{e^{\dfrac{K_\alpha \sqrt{Var(\hat{\theta})}}{\hat{\theta}}}}$$

$$U_\theta = \hat{\theta} e^{\dfrac{K_\alpha \sqrt{Var(\hat{\theta})}}{\hat{\theta}}}$$

where K_α is defined by

$$\alpha = \frac{1}{\sqrt{2\pi}} \int_{K_\alpha}^{\infty} e^{-\frac{t}{2}} dt = 1 - \Phi(K_\alpha)$$

$$Var(\hat{\theta}) = \left(\frac{\partial^2 \Lambda}{\partial \theta^2}\right)^{-1}$$

10.2 Preliminaries

The Higgins-Tsokos loss function is defined as follows:

$$L_{HT}(\hat{\theta},\theta) = \frac{f_1 e^{f_2(\hat{\theta}-\theta)} + f_2 e^{-f_1(\hat{\theta}-\theta)}}{f_1 + f_2} - 1, f_1, f_2 \succ 0.$$

We shall assume that θ behaves as a random variable that is characterized by the following Pareto probability density function:

$$f_1(\theta) = \frac{a}{b}\left(\frac{b}{\theta}\right)^{a+1} ; \theta \geq b \succ 0, a \succ 0.$$

The Pareto prior has been selected because of its mathematical tractability.

Using Exponential data, we will approximate the Pareto prior in such a way that good approximate Bayesian estimates of θ are obtained.

10.3 Main results

Let $x_1, x_2, \ldots, x_n$ denote the observations of a given system that are being characterized by the Exponential distribution.

We have the following posterior distribution:

$$h(\theta \backslash x) \frac{\theta^{n-a-1} e^{-\theta \sum_{1}^{n} x_i}}{\int_{b}^{\infty} \theta^{n-a-1} e^{-\theta \sum_{1}^{n} x_i} d\theta}, \theta \succ b..$$

Using the following approximate prior, along with the Higgins-Tsokos loss function, good approximate Bayesian estimates of θ are obtained:

$$f_1(\theta) = \frac{a}{b}\left(\frac{b}{\theta}\right)^{a+1} ; \theta \geq b \succ 0, a \succ 0.$$

$$a_0 = n, b_0 = \frac{n}{\sum_{1}^{n} x_i} - \frac{1}{f_1 + f_2} Ln\left(\frac{\sum_{1}^{n} x_i + f_2}{\sum_{1}^{n} x_i - f1}\right)$$

$$f_1 \prec \sum_{i=1}^{n} x_i$$

The corresponding Approximate Bayesian estimator of the parameter θ, subject to the above approximate prior is equal to

$$\frac{n}{\sum_{i=1}^{n} x_i}.$$

Considering the corresponding approximate posterior distribution, along with the equalities

$$P(\theta \succ L|\, x) = 1 - \alpha/2 \quad \text{and} \quad P(\theta \succ U|\, x) = \alpha/2,$$

we obtain the following lower and upper confidence bounds for θ:

$$L_{HT} \frac{n - Ln(1 - \alpha/2)}{\sum_{i=1}^{n} x_i} - \frac{1}{f_1 + f_2} Ln\left(\frac{\sum_{i=1}^{n} x_i + f_2}{\sum_{i=1}^{n} x_i - f_1}\right)$$

$$U_{HT} \frac{n - Ln(\alpha/2)}{\sum_{i=1}^{n} x_i} - \frac{1}{f_1 + f_2} Ln\left(\frac{\sum_{i=1}^{n} x_i + f_2}{\sum_{i=1}^{n} x_i - f_1}\right)$$

Hence the Approximate Bayesian confidence bounds for the Exponential population mean are

$$L_{b(HT)} = \frac{1}{\dfrac{n - Ln(\alpha/2)}{\sum_{i=1}^{n} x_i} - \dfrac{1}{f_1 + f_2} Ln\left(\dfrac{\sum_{i=1}^{n} x_i + f_2}{\sum_{i=1}^{n} x_i - f_1}\right)}$$

$$U_{b(HT)} = \frac{1}{\dfrac{n - Ln(1-\alpha/2)}{\sum_{i=1}^{n} x_i} - \dfrac{1}{f_1 + f_2} Ln\left(\dfrac{\sum_{i=1}^{n} x_i + f_2}{\sum_{i=1}^{n} x_i - f_1}\right)}$$

10.4 Numerical Results

To assess the performance of the above Approximate Bayesian confidence bounds, Approximate Bayesian confidence intervals will be constructed and compared to their classical counterparts corresponding to the Fisher Matrix bounds approach.

The widths of the Fisher Matrix bounds and the Approximate Bayesian confidence intervals are respectively denoted by l_F and l_{HT}.

For the Higgins-Tsokos loss function, we will consider f1=1, f2=1.

Example 1

These Exponential data, that we used in Chapter 10, were obtained by Washington State Department of Ecology that was conducting research on the amount of lead concentration in certain types of fish found in the Spokane River.

Lead (Pb) Concentrations in 1999 Spokane River Fish			
Source: WA State Dept. of Ecology report			
concentrations in parts per million (ppm)			
Filets	Trout	Whitefish	Sucker
	0.480	0.020	0.088
	0.071	0.020	0.210
	0.110	0.020	0.280
	0.320	0.020	0.030
	0.120	0.020	0.036
	0.220	0.065	0.047
	0.055	0.020	0.077
	0.320	0.037	0.069
	0.077	0.020	0.160
	0.081	0.036	0.088
	0.170		0.120
	0.130		0.054
	0.110		0.080
	0.081		0.059

	0.098		0.094
	0.180		0.059
	0.230		0.068
	0.082		0.020
	0.210		0.090
	0.200		0.046
	0.025		
	0.038		
mean	0.155	0.028	0.089
std dev	0.110	0.015	0.063

Table 4.1: Fisher Matrix bounds and Approximate Bayesian confidence intervals for the mean lead concentration in trout.

Confidence level	Fisher Matrix bounds	Approx.Bayesian bounds (HT)
80%	0.11791 - 0.20351	0.15301 – 0.16976
90%	0.10896 – 0.22021	0.14839 – 0.17022
95%	0.10199 – 0.23526	0.14404 – 0.17044
99%	0.08936 – 0.26851	0.13487 – 0.17061

Confidence level	$(l_F) \div (l_{HT})$
80%	5.1104

90%	5.0961
95%	5.0481
99%	5.0125

Table 4.2.: Fisher Matrix bounds and Approximate Bayesian confidence intervals for the mean lead concentration in whitefish.

Confidence level	Fisher Matrix bounds	Approx. Bayesian bounds (HT)
80%	0.01854 – 0.04167	0.02556 – 0.03204
90%	0.01649 – 0.04684	0.02403 – 0.03224
95%	0.01495 – 0.05166	0.02267 – 0.03234
99%	0.01229 – 0.06285	0.02004 – 0.03241

Confidence level	$(l_F) \div (l_{HT})$
80%	3.5694
90%	3.6967
95%	3.7962
99%	4.0873

Table 4.3: Fisher Matrix bounds and approximate Bayesian confidence intervals for the mean lead concentration in sucker.

Confidence level	Fisher Matrix bounds	Approx. Bayesian bounds (HT)
80%	0.06666 – 0.11816	0.08799 – 0.09875

90%	0.06136 – 0.12835	0.08507 – 0.09905
95%	0.05725 – 0.13756	0.08234 – 0.09919
99%	0.04984 – 0.15802	0.07662 – 0.09931

Confidence level	$(l_F) \div (l_{HT})$
80%	4.7862
90%	4.7918
95%	4.7661
99%	4.7677

Example 2

The following Exponential data represent a random sample of cycles to failure in ten-thousands for twenty heater switches subject to an overload voltage (Abdulaziz Elfessi & David M. Raineke,)2001

0.01, 0.034, 0.194, 0.567, 0.601, 0.712, 1.291, 1.367, 1.949, 2.37, 2.411, 2.875, 3.162, 3.28, 3.491, 3.686, 3.854, 4.211, 4.397, 6.473.

They conducted some studies on the above data and obtained the following maximum likelihood estimate and 95% confidence interval for the parameter θ: 0.4261 and (0.2603 , 0.6322)..

Table 4.4: Fisher Matrix bounds and approximate Bayesian confidence intervals of θ

Confidence level	Fisher Matrix bounds	*Approx.Bayesian bounds (HT)*
80%	0.32005 – 0.56732	0.38575 – 0.43256

90%	0.29464 – 0.61626	0.38459 – 0.44733
95%	0.27491 – 0.66049	0.38404 – 0.46210
99%	0.23932 – 0.75871	0.38361 – 0.49639

Confidence level	$(l_F) \div (l_{HT})$
80%	5.2824
90%	5.1262
95%	4.9395
99%	4.6053

Example 3

These Exponential data have been obtained by ReliaSoft Corporation (1998-2002). They assume that six identical units are reliability tested at the same application and operation stress levels. All the units under study fail during the test after operating for the following times (in hours):

96, 257, 498, 763, 1051, 1744.

Estimating the parameter graphically on probability plotting paper, they obtained the following estimate: 0.0012.

They also obtained the following maximum likelihood estimate: 0.00136.

Table 4.5: Fisher Matrix bounds and approximate Bayesian confidence intervals of θ

Confidence level	Fisher Matrix bounds	Approx.Bayesian bounds (HT)
80%	0.000806 – 0.00229	0.000931 – 0.00142
90%	0.000693 – 0.00266	0.000918 – 0.00158
95%	0.000611 – 0.00302	0.000912 – 0.00174
99%	0.000474 – 0.00390	0.000908 – 0.00210

Confidence level	$(l_F) \div (l_{HT})$
80%	*3.0347*
90%	*2.9712*
95%	*2.9094*
99%	*2.8741*

Example 4

Monte Carlo simulation has been used to generate the following 30 observations from the exponential distribution with mean equal to 9

2.0270, 4.0103, 30.0421, 0.1189, 2.7558. 13.7441, 13.3840, 27.0930, 7.3750, 3.7323, 23.4171, 0.06310. 5.6839, 8.7473, 10.2778, 25.2331, 10.1903, 0.3761, 3.3068, 3.4954, 6.9136, 1.8234, 16.3160, 2.4359, 19.9108, 2.5285, 3.9314, 3.4645, 6.9229, 10.4509.

Table 2: **Fisher Matrix bounds and Approximate Bayesian confidence intervals for the exponential population mean when the population mean is equal to 9.**

Confidence level	Fisher Matrix bounds	Approx.Bayesian bounds (HT)

80%	7.1184 – 11.3598	8.6182 – 9.2688
90%	6.6534 – 12.1537	8.4315 – 9.2861
95%	6.2873 – 12.8614	8.2527 – 9.2944
99%	5.6144 – 14.4028	7.8655 – 9.3009

Confidence level	$(l_F) \div (l_{HT})$
80%	6.5192
90%	6.4361
95%	6.3109
99%	6.1226

10.5 Summary and Conclusions

In this Chapter, we have derived Approximate Bayesian confidence bounds for the mean and the inverse of the mean of an Exponential distribution. The loss function that is employed is the Higgins-Tsokos loss function.

Based on the above numerical results we can conclude the following:

The obtained classical and Approximate Bayesian confidence intervals have great coverage accuracy. The Approximate Bayesian confidence intervals perform better; each of the obtained Approximate Bayesian confidence intervals contains the population mean and is strictly included in its classical counterpart.

Contrary to Fisher Matrix bounds method that uses the Z-table, the Approximate Bayesian approach and confidence bounds rely only on the observations that are under study.

With the Approximate Bayesian approach, Approximate Bayesian confidence intervals for the Exponential population mean are easily computed for any level of significance and any sample size..

Chapter 11 Approximate Bayesian Confidence Intervals For The shape parameter of a Gamma Distribution

11.1 Introduction

In Chapter 11, for large samples, we will derive closed-form confidence bounds for the shape parameter of the two-parameter Gamma distribution.

Considering the Square Error loss function, Approximate Bayesian confidence intervals for the shape parameter of a Gamma population will be constructed.

The two-parameter Gamma is defined as follows:

$$f(x) = \frac{1}{\Gamma(\alpha)\beta^{\alpha}} x^{\alpha-1} e^{-\frac{x}{\beta}}$$

$0 \prec x \prec \infty, \qquad \alpha \succ 0, \qquad \beta \succ 0.$

where α, is the shape parameter and β is the scale parameter; the inverse of the scale parameter, $1/\beta$, is the rate parameter.

The Gamma distribution is widely used to model positive and continuous variables having skewed distributions.

The following estimator is frequently used to obtain point estimates of the shape parameter corresponding to the two-parameter Gamma distribution:

$$\left(\frac{\bar{x}}{s}\right)^2$$

11.2 Preliminaries

Considering the Square Error Loss function, in Chapter 2 and Chapter 4, we obtained the following confidence bounds for the variance and the mean of a Normal distribution:

$$L_{\sigma^2(SE)} = \frac{\sum_{i=1}^{n}(x_i - \bar{x})^2}{n - 2 - 2\ln(\alpha/2)}$$

$$U_{\sigma^2(SE)} = \frac{\sum_{i=1}^{n}(x_i - \bar{x})^2}{n - 2 - 2\ln(1 - \alpha/2)}$$

$$U_{\mu(SE)} = \left(\frac{\sum_{i=1}^{n}(x_i - \bar{x})^2}{n-1} + \bar{x}^2 - \frac{\sum_{i=1}^{n}(x_i - \bar{x})^2}{n - 2 - 2\ln(\alpha/2)}\right)^{0.5}$$

$$L_{\mu(SE)} = \left(\frac{\sum_{i=1}^{n}(x_i - \bar{x})^2}{n-1} + \bar{x}^2 - \frac{\sum_{i=1}^{n}(x_i - \bar{x})^2}{n-2-2\ln(1-\alpha/2)} \right)^{0.5}$$

These models will be used to construct Approximate Bayesian confidence bounds for the shape parameter of the two-parameter Gamma distribution.

11.3 Main results

The Central Limit Theorem states the following: For large samples, the distribution corresponding to the sample mean $\bar{X}$ is approximately Normal, irrespective of the distribution of X.

Therefore, with large samples, to derive new Approximate Bayesian confidence bounds for the shape parameter of a Gamma distribution, we will use the Central Limit Theorem along with the above Approximate Bayesian confidence bounds for the variance and the mean of a Gaussian distribution.

Hence, using the Approximate Bayesian confidence bounds for the variance of a Normal distribution, we have the following:

$$\frac{\sum_{i=1}^{n}(\bar{x}-\mu)^2}{n-2-2\ln(\alpha/2)} \prec \frac{\sigma^2}{n} \prec \frac{\sum_{i=1}^{n}(\bar{x}-\mu)^2}{n-2-2\ln(1-\alpha/2)}$$

$$\frac{n\sigma^2}{n-2-2\ln(\alpha/2)} \prec \sigma^2 \prec \frac{n\sigma^2}{n-2-2\ln(1-\alpha/2)}$$

Thus, for large samples, we have the following Approximate Bayesian bounds for the population variance and mean of a Gamma distribution

Approximate Bayesian interval for the variance $\alpha\beta^2$: when the Gamma population mean is known.

$$L_{(SE)} = \frac{\sum_{i=1}^{n}(x_i - \mu)^2}{n - 2 - 2\ln(\alpha/2)}$$

$$U_{(SE)} = \frac{\sum_{i=1}^{n}(x_i - \mu)^2}{n - 2 - 2\ln(1 - \alpha/2)}$$

Approximate Bayesian interval for the variance $\alpha\beta^2$ when the Gamma population mean is not known.

$$L_{(SE)} = \frac{\sum_{i=1}^{n}(x_i - \bar{x})^2}{n - 2 - 2\ln(\alpha/2)}$$

$$U_{(SE)} = \frac{\sum_{i=1}^{n}(x_i - \bar{x})^2}{n - 2 - 2\ln(1 - \alpha/2)}$$

Approximate Bayesian interval for the population mean $\alpha\beta$

$$L_{\mu(SE)} = \left(s^2 + \bar{x}^2 - \frac{\sum_{i=1}^{n}(x_i - \bar{x})^2}{n - 2 - 2\ln(1 - \alpha/2)} \right)^{0.5}$$

$$U_{\mu(SE)} = \left(s^2 + \bar{x}^2 - \frac{\sum_{i=1}^{n}(x_i - \bar{x})^2}{n - 2 - 2\ln(\alpha/2)} \right)^{0.5}$$

Using these confidence intervals, we can easily derive the following confidence bounds for the inverse of the coefficient of variation of the Gamma distribution:

$$L = \sqrt{\frac{\left(s^2 + \bar{x}^2\right)\left(n - 2 - 2\ln(1 - \alpha/2)\right)}{\sum_{i=1}^{n}(x_i - \bar{x})^2} - 1}$$

$$U = \sqrt{\frac{\left(s^2 + \bar{x}^2\right)(n - 2 - 2\ln(\alpha/2))}{\sum_{i=1}^{n}(x_i - \bar{x})^2} - 1}$$

This gives us the following confidence bounds for the shape parameter of a Gamma distribution:

$$L_\alpha = \frac{\left(s^2 + \bar{x}^2\right)(n - 2 - 2\ln(1 - \alpha/2))}{\sum_{i=1}^{n}(x_i - \bar{x})^2} - 1$$

$$U_\alpha = \frac{\left(s^2 + \bar{x}^2\right)(n - 2 - 2\ln(\alpha/2))}{\sum_{i=1}^{n}(x_i - \bar{x})^2} - 1$$

11.4 Numerical Results

For the numerical results, we will use large Gamma data.

Example 1 SAS data

0.746, 0.357, 0.376, 0.327, 0.485, 1.741, 0.241, 0.777 0.768, 0.409,

0.252, 0.512, 0.534, 1.656, 0.742, 0.378, 0.714, 1.121, 0.597, 0.231,

0.541, 0.805, 0.682, 0.418, 0.506, 0.501, 0.247, 0.922, 0.880, 0.344, 0.519, 1.302, 0.275, 0.601, 0.388, 0.450, 0.845, 0.319, 0.486, 0.529, 1.547, 0.690, 0.676, 0.314, 0.736, 0.643, 0.483, 0.352, 0.636, 1.080.

$\bar{x} = 0.63362$

$S = 0.098336807$

SAS yielded the following point estimate of the shape and scale parameter:

$$: \left(\frac{\bar{x}}{s}\right)^2 = 4.082646$$

Table 1 Approximate Bayesian Confidence bounds for the shape parameter of the Gamma distribution

C.L. %	Confidence bounds for the shape parameter
80	4.000775 - 4.456600
90	3.989559 - 4.600397
95	3.984170 - 4.744194
99	3.979958 - 5.078080

Example 2 SAS data

620, 470, 260, 89, 388, 242, 103, 100, 39, 460, 284, 1285, 218, 393, 106, 158, 152, 477, 403, 103, 69, 158, 818, 947, 399, 1274, 32, 12, 134, 660, 548, 381, 203, 871, 193, 531, 317, 85, 1410, 250, 41, 1101, 32, 421, 32, 343, 376, 1512, 1792, 47, 95, 76, 515, 72, 1585, 253, 6, 860, 89, 1055, 537, 101, 385, 176, 11, 565, 164, 16, 1267, 352, 160, 195, 1279, 356, 751, 500, 803, 560, 151, 24, 689, 1119, 1733, 2194, 763, 555, 14, 45, 776, 1.

$\bar{x}$ =468.74444444

S = 475.927505248

Point estimate of shape parameter: 0.959264019

Table 2 **Approximate Bayesian Confidence bounds for the shape parameter of the Gamma distribution**

C.L. %	Confidence bounds for the shape parameter
80	0.941889 – 1.038629
90	0.939508 – 1.069147
95	0.938365 – 1.099665
99	0.937471 – 1.170526

Example 3 SAS data

1747, 945, 12, 1453, 14, 150, 20, 41, 35, 69, 195, 89,

1090, 1868, 294, 96, 618, 44, 142, 892, 1307, 310, 230, 30,

403, 860, 23, 406, 1054, 1935, 561, 348, 130, 13, 230, 250, 317, 304, 79, 1793, 536, 12, 9, 256, 201, 733, 510, 660, 122, 27, 273, 1231, 182, 289, 667, 761, 1096, 43, 44, 87, 405, 998, 1409, 61, 278, 407, 113, 25, 940, 28, 848, 41, 646, 575, 219, 303, 304, 38, 195, 1061, 174, 377, 388, 10, 246, 323, 198, 234, 39, 308, 55, 729, 813, 1216, 1618, 539, 6, 1566, 459, 946, 764, 794, 35, 181, 147, 116, 141, 19, 380, 609, 546.

$\overline{X}$ =459.513513514

S =477.755849874

Point estimate of the shape parameter: 0.925091198:

Table 3 **Approximate Bayesian Confidence bounds for the shape parameter of the Gamma distribution**

C.L. %	Confidence bounds for the shape parameter
80	0.9112782 – 0.988185
90	0.909386 – 1.012446
95	0.908477 – 1.036707
99	0.907766 – 1.093040

Exercise 4

Monte Carlo simulation has been used to generate the following 40 observations from a Gamma.

4.5046, 8.9119, 66.7603, 0.2643, 6.1241, 30.5425, 29.7423, 60.2067, 16.3891, 8.2941, 52.0380, 0.1402, 12.6309, 19.4385, 22.8395, 52.3378, 3.4389, 19.3268, 8.2350, 3.4737, 56.0736, 22.6451, 0.8359, 7.3484, 7.7675, 15.3635, 4.05222, 36.2578, 5.6189, 8.7365, 7.6990, 15.3844, 23.2242, 11.8542, 63.6975, 14.8772, 32.9585, 2.2127, 5,4132, 44.2462

$\bar{X}$ =20.297643

S = 19.3784360608

Point estimate of shape parameter: 1.097119087

Table 4 Approximate Bayesian Confidence bounds for the shape parameter of the Gamma distribution

C. L. %	Confidence bounds for the shape parameter
80	1.054678 – 1.290977
90	1.048863 – 1.365522
95	1.046070 – 1.440066
99	1.043886 – 1.613152

11.5 Summary and Conclusions

In Chapter 11, Approximate Bayesian confidence bounds for the shape parameter of a Gamma population have been constructed. The loss function that is employed is the Square Error loss function. Based on the above numerical results we can conclude the following:

The Approximate Bayesian confidence intervals have great coverage accuracy; the numerical results show that the obtained Approximate Bayesian confidence intervals contain the point estimates of the corresponding shape parameters.

The new Approximate Bayesian approach and closed-form confidence bounds rely only on the observations that are under study.

With the Approximate Bayesian approach, Approximate Bayesian confidence intervals for shape parameter of a Gamma population are easily obtained for any level of significance.

Chapter 12 Approximate Bayesian Confidence Intervals For The Scale and Rate parameters of a Gamma Distribution

12.1 Introduction

In this Chapter, we shall construct confidence intervals bounds for the scale and rate parameters of a two-parameter Gamma distribution.

Considering the Square Error Loss function, Approximate Bayesian confidence bounds for the scale and rate parameters of a Gamma distribution will be derived.

As mentioned in Chapter 11, the two-parameter Gamma is defined as follows

$$f(x) = \frac{1}{\Gamma(\alpha)\beta^{\alpha}} x^{\alpha-1} e^{-\frac{x}{\beta}}$$

$$0 < x < \infty, \quad \alpha > 0, \quad \beta > 0.$$

where α is the shape parameter and β is the scale parameter; the inverse of the scale parameter, $1/\beta$, is the rate parameter.

The following is frequently used to obtaining point estimates of the shape parameters corresponding to the two-parameter Gamma distribution:

$$\frac{s^2}{\bar{x}}$$

12.2 Preliminaries

In this Chapter, we will use the following confidence bounds for the variance and the shape parameter that are given in Chapter 11:

$$L_{\sigma^2(SE)} = \frac{\sum_{i=1}^{n}(x_i - \bar{x})^2}{n - 2 - 2\ln(\alpha/2)}$$

$$U_{\sigma^2(SE)} = \frac{\sum_{i=1}^{n}(x_i - \bar{x})^2}{n - 2 - 2\ln(1 - \alpha/2)}$$

$$L_{cv} = \frac{\left(s^2 + \bar{x}^2\right)(n - 2 - 2\ln(1 - \alpha/2))}{\sum_{i=1}^{n}(x_i - \bar{x})^2} - 1$$

$$U_{cv} = \frac{\left(s^2 + \bar{x}^2\right)(n - 2 - 2\ln(\alpha/2))}{\sum_{i=1}^{n}(x_i - \bar{x})^2} - 1$$

12.3 Main results

To derive the Approximate Bayesian confidence bounds for the scale parameter of the two-parameter Gamma distribution, we will use the Central limit along the obtained Approximate Bayesian confidence bounds for the variance and the shape parameter of the Gamma distribution.

Using the confidence bounds corresponding to the variance of the Gamma distribution

$$L_{\sigma^2(SE)} = \frac{\sum_{i=1}^{n}(x_i - \bar{x})^2}{n - 2 - 2\ln(\alpha/2)}$$

$$U_{\sigma^2(SE)} = \frac{\sum_{i=1}^{n}(x_i - \bar{x})^2}{n - 2 - 2\ln(1 - \alpha/2)}$$

along with the confidence bounds corresponding to the inverse of the shape parameter

$$L = \frac{\sum_{i=1}^{n}(x_i - \bar{x})^2}{\left(s^2 + \bar{x}^2\right)(n - 2 - 2\ln(\alpha/2)) - \sum_{i=1}^{n}(x_i - \bar{x})^2}$$

$$U = \frac{\sum_{i=1}^{n}(x_i - \bar{x})^2}{\left(s^2 + \bar{x}^2\right)(n - 2 - 2\ln(1 - \alpha/2)) - \sum_{i=1}^{n}(x_i - \bar{x})^2}$$

we obtain the following confidence bounds for the square of the scale parameter:

$$L = \frac{\sum_{i=1}^{n}(x_i - \bar{x})^2}{n - 2 - 2\ln(\alpha/2)} \left(\frac{\sum_{i=1}^{n}(x_i - \bar{x})^2}{\left(s^2 + \bar{x}^2\right)\left(n - 2 - 2\ln(\alpha/2)\right) - \sum_{i=1}^{n}(x_i - \bar{x})^2} \right)$$

$$U = \frac{\sum_{i=1}^{n}(x_i - \bar{x})^2}{n - 2 - 2\ln(1 - \alpha/2)} \left(\frac{\sum_{i=1}^{n}(x_i - \bar{x})^2}{\left(s^2 + \bar{x}^2\right)\left(n - 2 - 2\ln(1 - \alpha/2)\right) - \sum_{i=1}^{n}(x_i - \bar{x})^2} \right)$$

Hence, we have the following confidence bounds for the scale parameter of a The Gamma distribution:

$$L = \frac{\sum_{i=1}^{n}(x_i - \bar{x})^2}{\sqrt{\left(s^2 + \bar{x}^2\right)(n - 2 - 2\ln(\alpha/2))^2 - (n - 2 - 2\ln(\alpha/2))\sum_{i=1}^{n}(x_i - \bar{x})^2}}$$

$$U = \frac{\sum_{i=1}^{n}(x_i - \bar{x})^2}{\sqrt{\left(s^2 + \bar{x}^2\right)(n - 2 - 2\ln(1 - \alpha/2))^2 - (n - 2 - 2\ln(1 - \alpha/2))\sum_{i=1}^{n}(x_i - \bar{x})^2}}$$

Therefore, the confidence bounds for the rate parameter of a Gamma distribution are the following::

$$L = \frac{\sqrt{\left(s^2 + \bar{x}^2\right)(n - 2 - 2\ln(1 - \alpha/2))^2 - (n - 2 - 2\ln(1 - \alpha/2))\sum_{i=1}^{n}(x_i - \bar{x})^2}}{\sum_{i=1}^{n}(x_i - \bar{x})^2}$$

$$U = \frac{\sqrt{\left(s^2 + \bar{x}^2\right)(n - 2 - 2\ln\alpha/2))^2 - (n - 2 - 2\ln(\alpha/2))\sum_{i=1}^{n}(x_i - \bar{x})^2}}{\sum_{i=1}^{n}(x_i - \bar{x})^2}$$

12.4 Numerical Results

For the numerical results, we will use large Gamma data.

Example 1 SAS data

0.746 0.357 0.376 0.327 0.485 1.741 0.241 0.777 0.768 0.409

0.252 0.512 0.534 1.656 0.742 0.378 0.714 1.121 0.597 0.231

0.541 0.805 0.682 0.418 0.506 0.501 0.247 0.922 0.880 0.344

0.519 1.302 0.275 0.601 0.388 0.450 0.845 0.319 0.486 0.529

1.547 0.690 0.676 0.314 0.736 0.643 0.483 0.352 0.636 1.080

$\overline{x} = 0.63362$

$S = 0.098336807$

SAS yields the following estimate of the shape and scale parameters:

Estimate of the scale parameter: $\dfrac{s^2}{\overline{x}} = 0.155198$

Estimate of the rate parameter: 6.443382

Table 1: Confidence bounds for the scale parameter of the Gamma distribution

C.L. %	Confidence bounds for the scale parameter

80	0.143864 – 0.158056
90	0.139282 – 0.158456
95	0.135428 – 0.158649
99	0.127254 - 0.158800

Table 2 Confidence bounds for the rate parameter of the Gamma distribution

C. L. %	Confidence bounds for the rate parameter
80	6.326855 – 6.975253
90	6.310889 – 7.179664
95	6.303220 – 7.385110
99	6.297224 – 7.858325

Example 2 SAS data

620, 470, 260, 89 , 388, 242, 103, 100, 39, 460 , 284, 1285, 218, 393, 106, 158, 152, 477, 403, 103, 69 , 158 ,818, 947, 399, 1274, 32 , 12 ,134 , 660, 548 ,381 ,203 , 871, 193, 531, 317, 85 , 1410, 250 ,41 , 1101, 32, 421, 32 , 343, 376 , 1512, 1792, 47, 95, 76, 515, 72,1585, 253, 6, 860, 89 , 1055, 537, 101, 385, 176, 11, 565, 164, 16, 1267, 352, 160, 195, 1279, 356, 751, 500, 803, 560, 151, 24, 689, 1119, 1733, 2194, 763, 555, 14, 45, 776, 1.

$\overline{x}$ =468.74444444

S = 475.927505248

Estimate of scale parameter: 488.650085

Estimate of the rate parameter: 0.002046454

Table 3: Confidence bounds for the scale parameter of the Gamma distribution

C. L. %	Confidence bounds for the scale parameter
80	460.377645 - 495.337918-
90	450.400759 – 496.269320
95	440.867642 – 496.718142
99	420.281633 – 497.06955

Table 4: Confidence bounds for the rate parameter of the Gamma distribution

C. L. %	Confidence bounds for the rate parameter
80	0.0020188 – 0.0021721
90	0.0020150 - 0.0022202
95	0.0020132 – 0.0022683
99	0.0020180 – 0.0023794

Example 3 SAS data

1747, 945, 12, 1453, 14, 150, 20, 41, 35, 69, 195, 89,

1090, 1868, 294, 96, 618, 44, 142, 892, 1307, 310, 230, 30,

403, 860, 23, 406, 1054, 1935, 561, 348, 130, 13, 230, 250,

317, 304, 79, 1793, 536, 12, 9, 256, 201, 733, 510, 660,

122, 27, 273, 1231, 182, 289, 667, 761, 1096, 43, 44, 87,

405, 998, 1409, 61, 278, 407, 113, 25, 940, 28, 848, 41, 646, 575, 219, 303, 304, 38, 195, 1061, 174, 377, 388, 10, 246, 323, 198, 234, 39, 308, 55, 729, 813, 1216, 1618, 539, 6, 1566, 459, 946, 764, 794, 35, 181, 147, 116, 141, 19, 380, 609, 546.

$\overline{X}$ = 459.513513514

S = 477.755849874

Estimate of the scale parameter: 496.7223931

Estimate of the rate parameter: 0.002013197

Table 5: Confidence bounds for the scale parameter of the Gamma distribution

C.L. %	Confidence bounds for the scale parameter
80	472.916267 – 502.278094
90	464.390820 – 503.049553
95	456.183203 – 503.421082
99	438.252898 – 503.711915

Table 6: Confidence bounds for the rate parameter of the Gamma distribution

C.L.	Confidence bounds for the rate parameter

%	
80	0.00199093 – 0.00211454
90	0.00198788 – 0.00215336
95	0.00198641 – 0.00219210
99	0.00198526 – 0.00228179

Exercise 4

Monte Carlo simulation has been used to generate the following 40 observations from a Gamma distribution

4.5046, 8.9119, 66.7603, 0.2643, 6.1241, 30.5425, 29.7423, 60.2067, 16.3891, 8.2941, 52.0380, 0.1402, 12.6309, 19.4385, 22.8395, 52.3378, 3.4389, 19.3268, 8.2350, 3.4737, 56.0736, 22.6451, 0.8359, 7.3484, 7.7675, 15.3635, 4.05222, 36.2578, 5.6189, 8.7365, 7.6990, 15.3844, 23.2242, 11.8542, 63.6975, 14.8772, 32.9585, 2.2127, 5,4132, 44.2462

$\overline{X} = 20.297643$

$S = 19.3784360608$

Estimate of the scale parameter: 18.50085668

Estimate of the rate parameter: 0.054051551

Table 7: Confidence bounds for the scale parameter of the Gamma distribution

C.L.	Confidence bounds for the scale parameter

%	
80	16.317761 – 19.063310
90	15.614121 – 19.143194
95	14.970570 – 19.181819
99	13.668167 – 19.212127

Table 8: Confidence bounds for the rate parameter of the Gamma distribution

C. L. %	Confidence bounds for the rate parameter
80	0.05245676 - 0.06128292
90	0.05223789 – 0.06404459
95	0.05213270 – 0.06679772
99	0.05205046 – 0.07316270

12.5 Summary and Conclusions

In Chapter12, closed-form Approximate Bayesian confidence bounds for the scale and rate parameters of a Gamma population have been derived. The loss function that is employed is the Square Error loss function. Based on the above numerical results we can conclude the following:

The Approximate Bayesian confidence bounds for the Gamma scale and rate parameters have great coverage accuracy.

The new Approximate Bayesian approach and models rely only on the observations that are under study.

With the new Approximate Bayesian approach, confidence intervals for the scale and rate parameters of a Gamma population are easily constructed for any level of significance.

References

Camara, Vincent A. R. (2003) "Approximate Bayesian Confidence Intervals For The Variance Of A Gaussian Distribution," *Journal of Modern Applied Statistical Methods*: Vol. 2: Iss. 2, Article 8.

Camara, Vincent A. R. (2007) "Approximate Bayesian Confidence Intervals for the Mean of an Exponential Distribution Versus Fisher Matrix Bounds Models," Journal of Modern Applied Statistical Methods: Vol. 6: Iss. 1, Article 14.

Camara, Vincent A. R. (2009) "A New Approximate Bayesian Approach for Decision Making About the Variance of a Gaussian Distribution Versus the Classical Approach," *Journal of Modern Applied Statistical Methods*: Vol. 8: Iss. 1, Article 22.

Camara, Vincent A. R. (2009) "Approximate Bayesian Confidence Intervals for The Mean of a Gaussian Distribution Versus Bayesian Models," *Journal of Modern Applied Statistical Methods*: Vol. 8: Iss. 2, Article 17.

Vincent A. R. Camara. New Approximate Bayesian Confidence. Intervals for the Coefficient of Variation of a Normal Distribution May, 2012 Vol. 11, No. 1 - pages167 – 178 JMASM.com

M. Fogel, The Statistics Problem Solver,p.p. 502-505, 1991

Miller S. G. (1991), Asymptotic test statistics for Coefficient of variation, Comm. Stat. Theory and methods 20(10), 3351-3363.

Nelson, Wayne, *Applied Life Data Analysis*, John Wiley & Sons, Inc., New York, 1982.

Robert R. Britney and Robert L. Winkler, Bayesian III point estimation under various loss functions, Proc. Business and Economic Statistics Section, Amer. Statist.

Assoc., 1968, pp. 356-364.

Ronald V. Canfield, A Bayesian approach to reliability estimation using a loss function, IEEE Trans. Reliability R-19(1):13-16 (1970).

Vincent A. R. Camara and C. P. Tsokos, Sensitivity Behavior of Bayesian Reliability Analysis for different Loss Functions, International Journal of Applied Mathematics,

November (1999).

Bernard Harris, A survey of statistical methods in system reliability using Bernoulli ampling of components, in Proceedings of the conference on the theory and

applications of Reliability with emphasis on Bayesian and Nonparametric Methods, Academic, New York, (1976).

J. J. Higgins and C. P.Tsokos, Comparison of Bayes estimates of failure intensity for fitted priors of life data, in Proceedings of the Conference on the Theory and

Applications of Reliability with Emphasis on Bayesian and Nonparametric Methods, Academic, New York, (1976).

J. J. Higgins and C. P. Tsokos, On the behavior of some quantities used in Bayesian reliability demonstration tests, IEEE Trans. Reliability R-25(4):261-264(1976).

Vincent A. R. Camara and C. P. Tsokos, The effect of Loss Functions on Empirical Bayes Reliability Analysis, Journal of Engineering Problems, (1999).

R. E. Schafer et al., Bayesian reliability demonstration, PhaseI- Data for the a priori distribution, Rome Air Development Center, Griffis AFBNY RADC-TR-69-389.

(1970).

R. E. Schafer et al., Bayesian reliability, Phase II- Development of a priori distribution. Rome Air Development Center, Griffis AFR, NY RADC-YR-71-209,

(1971)

Vincent A. R. Camara and Chris P. Tsokos Bayesian, Reliability Modeling with a new Loss Function Function, STATISTICA, July (1999).

R. E. Schafer et al ., Bayesian reliability demonstration Phase III - Development of test plans, Rome Air development Center, Griffs AFB, NY RADC-TR-73-39,

(1973).

R. E. Shafer and Anthony J. Feduccia, Prior distribution fitted to observed reliability data, IEEE Trans. Reliability R-21 (3):148-154 (1972).

Bhattacharya, S. K. Bayesian Approach to Life Testing and Reliability Estimation. Jour. Amr. Statist. Assoc. , 62, 48-62, (1967)

Vincent A. R. Camara and Chris P. Tsokos. Bayesian Estimate of a Parameter and Choice of the Loss Function. Nonlinear Studies journal, (1999).

Drake, A. W., Bayesian Statistics for the Reliability Engineer. Proc. 1966 Annual Symposium On Reliability, 315-320, (1966).

Vincent A. R. Camara and C. P. Tsokos, Effect of Loss Functions on Bayesian Reliability Analysis. Proceedings of Int. Conf. On Nonlinear Problems in Aviation and Aerospace, pages 75-90, (1996).Winkler Robert L., Introduction to Bayesian Inference and Decision making, pages 174-

179, 395-397 ; (1972).

Prem S. Mann, Introductory Statistics, Third edition, page 504, (1998).

James T. McClave/Terry Sincich A first course in Statistics, page 301, Sixth edition, (1997).

Drake, A. W. Bayesian Statistics for the Reliability Engineer. Proc.1966 Annual Symposium on Reliability, p.p. 315-320, 1966.

Vincent A. R. Camara and C. P. Tsokos, Bayesian Reliability Modeling with Applications, UMI Publishing Company, and Registered at the United States Library of

Congress, 1998.

VilTA

Vincent A. R. Camara is a Mathematics/Statistics educator and researcher. He earned a Doctorate in Mathematics/Statistics, a Master's degree in Pure Mathematics, a Master's degree in Applied Mathematics and a Bachelor's degree in Mathematical science.

He is a member of Phi Kappa Phi and Pi Mu Epsilon.

He is featured in Marquis Who's Who in America and in Marquis Who's Who in the World.

www.ingramcontent.com/pod-product-compliance
Lightning Source LLC
Chambersburg PA
CBHW080253180526
45167CB00006B/2513